Handbook of Nonlinear Regression Models

STATISTICS: Textbooks and Monographs

A Series Edited by

D. B. Owen, Coordinating Editor
Department of Statistics
Southern Methodist University
Dallas, Texas

R. G. Cornell, Associate Editor
for Biostatistics
University of Michigan

W. J. Kennedy, Associate Editor
for Statistical Computing
Iowa State University

A. M. Kshirsagar, Associate Editor
for Multivariate Analysis and
Experimental Design
University of Michigan

E. G. Schilling, Associate Editor
for Statistical Quality Control
Rochester Institute of Technology

ADDITIONAL VOLUMES IN PREPARATION

Handbook of Nonlinear Regression Models

DAVID A. RATKOWSKY

Program in Statistics
Washington State University
Pullman, Washington

MARCEL DEKKER, INC.   New York and Basel

Library of Congress Cataloging-in-Publication Data

Ratkowsky, David A.
  Handbook of nonlinear regression models / David A. Ratkowsky
      p.   cm. -- (Statistics, textbooks and monographs ; v. 107)
  Includes bibliographical references.
  ISBN 0-8247-8189-9 (alk. paper)
  1. Regression analysis. 2. Nonlinear theories. 3. Parameter
estimation. 4. Linear models (Statistics) I. Title. II. Series.
  QA278.2.R369   1989
  519.5'36–dc20   89-27493
                                                            CIP

This book is printed on acid-free paper.

MARCEL DEKKER, INC.
270 Madison Avenue, New York, New York 10016

Current printing (last digit):
10 9 8 7 6 5 4 3 2 1

PRINTED IN THE UNITED STATES OF AMERICA

Preface

The purpose of this handbook is to bring together in one volume a wide range of commonly used nonlinear regression models and to describe the statistical properties of the estimators of the parameters of those models. The underlying philosophy of the approach has already been described in detail in my earlier book *Nonlinear Regression Modeling: A Unified Practical Approach*, Marcel Dekker, 1983. In that work, I emphasized the desirability of selecting nonlinear regression models that are close-to-linear in their estimation behavior. A close-to-linear nonlinear regression model is one whose parameter estimators are close to being unbiased, normally distributed, minimum variance estimators, a condition that always applies in linear regression models but is approximated to a greater or lesser degree in nonlinear regression models, depending on the model and the sample size. Properties of close-to-linear models are described in detail in Chapter 2.

This handbook will attempt to examine all commonly used nonlinear regression models and to indicate, using an asterisk (∗) to denote reasonably close-to-linear estimation behavior and a double asterisk (∗∗) very close-to-linear estimation behavior, which basic models, or reparameterizations of the basic models, should be used in practice. A concomitant benefit of a close-to-linear model is that convergence to the least-squares estimates is likely to occur, and to occur rapidly, from a set of reasonable starting values using standard algorithms such as the Gauss–Newton method or the Newton–Raphson method.

Although this handbook is not primarily concerned with how the various nonlinear regression models have arisen, some attention will be paid, for some models, to the differential equations for which those models are solutions. Particularly helpful, in this area, is the approach of K. E. F. Watt (1968), who developed differential equations by considering the presence or absence of five factors in combination, that is, whether the derivative dY/dX is linear in X, proportional to Y, inversely proportional to X, approaches zero as Y approaches an asymptote, or approaches infinity as X approaches some minimum value $X_{\min}$. The solutions of the 32 differential equations thus formed include some of the very valuable commonly used nonlinear regression models, such as the logistic and the asymptotic regression models. Of course, this is not the only way of obtaining nonlinear regression models, but it is a logical and systematic approach to the subject.

The handbook is divided into two parts. Part I consists of three chapters and contains the background to, and philosophy of, nonlinear regression modeling, together with material on what a regression analysis is, the contrast between linear and nonlinear regression analysis, and what things are necessary to consider in regression analyses. Chapter 3 details, by means of an illustrative example, many aspects of regression analysis. Part II of the handbook deals with the models themselves. The question arose as to how the various models ought to be arranged in this handbook, whether by the purpose or use to which they are put, or by their shape, or by the number of explanatory (regressor) variables they contain. It was decided that shape of curve, within a category determined by the number of explanatory variables, was the only sensible criterion of classification, since a given curve may be used for quite different purposes in different applied disciplines. Given this decision, Chapters 4–6 deal with models with a single explanatory variable. Chapter 4 is devoted to ascending or descending convex/concave curves, that is, ones without maxima, minima, or inflection points. Chapter 5 deals with models having an inflection point but without a maximum or a minimum, although they may have asymptotes. Chapter 6 considers models with maxima and/or minima and that may also have inflection points. Chapter 7 deals with models having more than one explanatory variable. Chapter 8 deals with other nonlinear regression models that do not fall into the headings of Chapters 4–7 and also some others that have been excluded from serious consideration in this handbook. Chapter 9 considers the question of how to obtain initial parameter estimates as a starting point for finding the least-squares estimates. Chapter 10 is a summary of the handbook that attempts to present a few unifying concepts.

Naturally, it would be impossible to examine every nonlinear regression model that has ever been used (or could be used) in practice. Hence, attention has been restricted to models containing a maximum of six parameters. Most models that satisfy the requirement of being close-to-linear will have fewer parameters than this, usually a maximum of four. It is very rare for a model with more than four parameters to possess good estimation properties, unless that model fits the data extremely well. Nevertheless, there are likely to be omissions in this handbook. The author apologizes to readers who fail to find their favorite models here. Every attempt was made to seek out the commonly used models, but it is inevitable that some may have slipped through the net.

My earlier book dealt primarily with the properties of the estimators and developed a philosophy that insisted on models for which the estimators have properties closely approaching those of the estimators of parameters in linear regression models. This handbook is an attempt to put that philosophy into practice by providing modelers with the information they need to select models for practical use. Of course, the methodology described and presented in my earlier book and in Part I of this handbook, and the computer software presented in the appendices of that book and this handbook, will enable the modeler to evaluate the estimation behavior of any nonlinear regression model. However, to use the software, one must derive and code the first and second derivatives of each model of interest with respect to the parameters. This is a tedious, time-consuming task, which requires knowledge of the rules of differentiation that may be outside the capabilities of many modelers. It is hoped that this handbook will relieve modelers of having to perform that task by providing the results of such evaluations for most commonly used models.

Of the various people who contributed to the realization of this handbook, I would particularly like to thank Dr. Phil West, of the CSIRO Division of Forestry and Forest Products, Hobart, Tasmania, Australia, who, at an early stage, recognized the need for a handbook on nonlinear regression models and encouraged me to write one. Subsequently, he read critically the material I had written for Part I of this handbook, and his detailed comments helped shape its final form. It was particularly valuable to me to have had an experienced user of nonlinear regression models viewing the material from a user's rather than from a statistician's point of view. The bulk of the typing was done by Anne Greenhalgh of Hobart, Tasmania; Eliane Yochum of Santa Barbara, California, typed a portion of the revised manuscript. Any errors herein are my sole responsibility and fault. I thank my former employers, the CSIRO, Australia, and the University of California at Santa Barbara, for providing me with the encouragement and the facilities to undertake and complete this project.

David A. Ratkowsky

Contents

Handbook of Nonlinear Regression Models

I

The Background

This handbook is divided into two parts. Part I deals with the background to nonlinear regression modeling and Part II with the models themselves.

Part I has three chapters. The first of these introduces the reader to regression modeling in general, describes the difference between regression models and functional or structural relationships, and details the various considerations that one must undertake in setting out on a modeling task. Chapter 2 deals with nonlinear regression models, describing (1) how they differ from linear regression models, (2) the various measures of nonlinear behavior of the estimators of the parameters, (3) some basic principles of reparameterization (to produce a parameterization with better statistical behavior), (4) construction of confidence regions and confidence intervals, and (5) some fallacies, misconceptions, and traps in nonlinear regression modeling. Chapter 3 serves to illustrate by example many of the principles of nonlinear regression models by confining attention to a particular class of linear and nonlinear regression models, that of relating moisture content to water activity in foodstuffs.

1

An Introduction to Regression Modeling

1.1 REGRESSION MODELS

A simple example of a regression model is the straight-line relationship between a pair of variables denoted X and Y. If the relationship between the two variables were exact, as determined by some physical law, one would write

$$Y = \alpha + \beta X \tag{1.1.1}$$

where α and β are called the *parameters*, which are known physical constants linking X and Y. However, in applied science, it is customary not to have exact physical relationships, but empirically determined ones, between observed variables. For example, with a set of n observations of X and Y, and the belief that X and Y are related by a simple straight-line relationship, one could write

$$Y_t = \alpha + \beta X_t + \epsilon_t \tag{1.1.2}$$

where Y_t represents the *response* variable, X_t represents the *explanatory* variable, and ϵ_t represents the *error* term at the tth data point ($t = 1,2,\ldots,n$). The parameters α and β are not known physical constants as in (1.1.1) but must be estimated from the data set.

In a regression model such as (1.1.2), it is customary to assume that the explanatory variable X_t, also referred to as the *regressor*, *predictor*, or *independent* variable, is preset or is measured without error. This assumption is tenable when the values of X_t are fixed or controlled experimental settings or when they are observed values measured without error. When X_t does not satisfy this requirement, more complicated models arise. These will be discussed in Section 1.3.

The response variable Y_t, also referred to as the *dependent* variable, is, unlike X_t, not controlled or preset and therefore is not completely determined by the function $\alpha + \beta X_t$. Hence the need for the error term ϵ_t, also called the *stochastic* term, in order to explain Y_t completely.

One can generalize, from the straight-line regression model to other regression models, and write

$$Y_t = f(X_t; \boldsymbol{\theta}) + \epsilon_t \tag{1.1.3}$$

Y_t, X_t, and ϵ_t are the same as in (1.1.2), but $f(X_t; \boldsymbol{\theta})$ is mathematical symbolic language for describing a function of the explanatory variable X_t and the parameters $\boldsymbol{\theta}$, the latter indicating a vector $[\theta_1, \theta_2, \ldots, \theta_p]$ of p parameters. In (1.1.2), $p = 2$, and α and β take the place of θ_1 and θ_2. The explanatory variable X_t should be written more properly as a vector $\mathbf{X}_t$, since there may be more than one explanatory variable, as in

$$Y_t = \alpha + \beta X_{1t} + \gamma X_{2t} + \epsilon_t \tag{1.1.4}$$

However, as most models in this handbook have only a single explanatory variable, the notation X_t has been chosen for simplicity.

It is customary for statisticians to denote parameters by using Greek letters such as α, β, γ, θ_1, θ_2, but other practitioners use capital Roman letters such as A, B, C, D or even lowercase Roman letters such as a, b, c, d.

1.2 LINEAR VERSUS NONLINEAR REGRESSION MODELS

Models such as (1.1.2)–(1.1.4) are called *linear* regression models because, in each case, all their parameters appear linearly. Some textbooks may refer to a model such as

$$Y_t = \alpha + \beta X_t + \gamma X_t^2 + \epsilon_t \qquad (1.2.1)$$

as a nonlinear regression model because Y_t is nonlinear in X_t. That is not the sense in which *nonlinear* is used in this handbook. If one wishes to emphasize the relationship between the response variable and the explanatory variable, (1.2.1) is better referred to as a *curvilinear* regression model. In the sense of this handbook, a nonlinear regression model is one in which at least one of its parameters appears nonlinearly, for example,

$$Y_t = \alpha X_t^\beta + \epsilon_t \qquad (1.2.2)$$

In the formal sense, nonlinear means that at least one of the derivatives of Y_t with respect to α and β is a function of at least one of those parameters. In (1.2.2), the derivative of Y_t with respect to α and the derivative of Y_t with respect to β are both functions of α and/or β, so that this model is a nonlinear regression model.

There are two main reasons why linear regression models are preferred to nonlinear regression models. The first reason is that linear regression is mathematically easier, with the estimators of the parameters in linear models being obtained from an explicit mathematical expression. For nonlinear regression models, one must use either an iterative procedure employing a mathematical algorithm or some exhaustive search procedure, the latter being highly inefficient of computational time. Further discussion of this appears in Chapter 2.

The second main reason for the popularity of linear regression models stems from the fact that often the investigator does not know what is an appropriate model for the set of data in hand. Sometimes, a large number of potential explanatory variables, X_{1t}, X_{2t}, X_{3t}, ..., etc., are measured, and it is not known which of these are useful for explaining a large part of the variation in the response variable Y_t. Often, the investigator will use the technique called *multiple linear regression* in an attempt to obtain a regression equation with a small number of explanatory variables that will explain a considerable proportion of the variation in Y_t. Model (1.1.4) is the simplest case of a multiple linear regression model, one with just two explanatory variables, X_{1t} and X_{2t}. Draper and Smith (1981) devote a chapter to methods for selecting the "best" regression equation of this multiple regression type. It should be clear, however, that even the best regression equation in this sense cannot be any more than a gross approximation to the true underlying model, which is almost certainly a nonlinear regression model. Multiple variable regression is a subject fraught with difficulties, since the set of explanatory variables is usually highly correlated, often resulting in "multicollinearity" and biased regression estimators (Gunst and Mason, 1980; Miller, 1984).

Nonlinear regression models with more than one explanatory variable, with few exceptions, such as the following, for example,

$$Y_t = X_{1t}X_{2t}/(\alpha + \beta X_{1t} + \gamma X_{2t} + \delta X_{1t}X_{2t}) + \epsilon_t \qquad (1.2.3)$$

tend to be very complicated algebraically and hence are not as common as nonlinear regression models with only a single explanatory variable. Most of the models in this handbook (Chaps. 4–6) contain only one explanatory variable, but one chapter (Chap.

7) is devoted to models with more than a single explanatory variable, all of these of a relatively simple type. As will be seen in Chapter 2, complicated nonlinear regression models lead not only to difficulties in the estimation of their parameters but also to undesirable statistical properties of their estimators.

1.3 FUNCTIONAL AND STRUCTURAL RELATIONSHIPS

In the standard regression model (1.1.3), the explanatory variable X_t, in contrast to the response variable Y_t, is assumed to be free of error. This assumption is tenable when the values of X_t are predetermined, fixed, or controlled experimental settings or when they are observed values measured without error. Often in practice, however, X_t does not satisfy this requirement and, like Y_t, has a component of error or reflects biological or physical variability. This situation gives rise to more complicated regression models known as functional or structural relationships. However, except for the very special case of the simple, straight-line regression model (1.1.1), statistical solutions to the functional relationship problem are not straightforward. Hence, there is a tremendous temptation for practitioners wanting to describe their results by using mathematical relationships to ignore the fact the X_t really contains a component of error. This desire is understandable and cannot be condemned, but it should be noted that as the extent to which the error in X_t increases, so also does the unreliability of the estimators of the parameters in the model. Stated differently, (1.1.3) contains certain assumptions, and the validity of the conclusions obtained depends on the validity of the assumptions. The more the latter are violated, the less reliable are the conclusions. Many regression relationships derived in practice use measured variables for X_t and Y_t. That is, users often are interested in a particular variable, which is thereby called Y_t, and they wish to find other variables that will explain Y_t in a regression relationship. If these other variables are subject to a similar degree of measurement error or random variation, as is Y_t, the derived relationship could be grossly in error. Therefore, for both linear and nonlinear regression models, the experimenter must attempt to make X_t as close as possible to a predetermined or controlled variable.

Although the regression model (1.1.3) requires the explanatory variable X_t to be without error, this is often not possible in practice, as many data sets consist of a set of variables, all of which may actually be measurements made on individuals drawn from some population. Consider, for example, samples of a species of squid caught by a scientific research vessel. The researchers may wish to find some relationship between the squid's body weight and the thickness of its mantle. Assume that there is a simple linear relationship between the cube root of the body weight and mantle thickness. This is close to being realistic for a species of squid caught in Bass Strait, the body of water that separates Tasmania from mainland Australia. Let us denote the cube root of the body weight by Y_t, which becomes the response variable of interest, and the mantle thickness by X_t. Suppose that the researchers measure the mantle thickness as accurately as they can, bearing in mind that the tissue of animals is extensible and that the thickness may depend strongly on whether the animal is alive, is in rigor, or has been dead for some considerable time. One may ask whether the measurement error in X_t is the only source of nonconstancy of the explanatory variable? The answer is no. The reason is that the shape of squid is not identical from individual to individual, so that squid with the same body weight can have different mantle thicknesses and, conversely, squid with the same mantle thicknesses can have different body weights. The explanatory variable, mantle thickness, is not a controlled variable here; instead, we have a case of *joint* variation of the variates X_t and Y_t.

The preceding case is different from the problem in which there may be a linear relationship between the yield Y_t of a chemical reaction and the temperature of the reaction chamber X_t. In this case, the temperature is restricted to certain fixed values under the control of the experimenter, and Y is measured at each of those values. If a further set of readings is taken using the same values of X_t, one must expect to get a different set of responses

Y_t. In other words, Y_t is subject to error, but X_t is not. This is the usual regression model, and (1.1.3) is an example of it. The assumption that Y_t is the only variable containing error is tenable in the chemical reaction example but not in the squid example in the preceding paragraph. It must be emphasized that relationships obtained using the standard regression model when, in fact, there is joint variation of the response and explanatory variables may be grossly in error. A good discussion of the various forms of straight-line models when there is a single explanatory variable is given by Ricker (1973). However, the solution recommended by Ricker for many practical problems—to use the geometric mean regression (or standard major axis) as the estimate for the slope of the relationship—is controversial (Jolicoeur, 1975; Sprent and Dolby, 1980). Since no solution is available for problems with more than one explanatory variable, or for nonlinear relationships, there is, in practice, no real alternative to the standard regression model (1.1.3). Thus, even when the assumption of the fixedness of X_t is not reasonable, users may be forced to adopt that model anyway. Use of (1.1.3) when it is not appropriate is widespread in the scientific world. The consequences of ignoring the variation in X_t, however, may be considerable if the variation in X_t is large.

1.4 CONSIDERATIONS IN REGRESSION ANALYSIS

One might reasonably ask how a modeler commences a regression task. If there is some precedent for the use of a certain model, which may be a nonlinear regression model, resulting from some underlying theory or from a large base of empirical evidence, the modeler can confidently use that model, although the choice of parameterization is still open to question. To understand what is meant by the word *parameterization*, consider the following two models:

$$Y_t = \alpha X_t/(X_t + \beta) + \epsilon_t \tag{1.4.1}$$

and

$$Y_t = X_t/(\theta_1 X_t + \theta_2) + \epsilon_t \qquad\qquad (1.4.2)$$

Models (1.4.1) and (1.4.2) are the same *basic* model. If one defines $\theta_1 = 1/\alpha$ and $\theta_2 = \beta/\alpha$, then the two models (parameterizations) produce identical values of Y_t for the same value X_t. As will be seen in Chapter 2, however, the statistical properties of the estimators of the parameters in one or these models may be much better (in the sense that will be described in Chap. 2) than the statistical properties of the estimators in the other model.

If there is no precedent for the use of a particular basic model and if there is only a single explanatory variable, the modeler may try to match up a plot of the data of Y_t versus X_t against the graphs of the various models given in Chapters 4–6 in an attempt to find one that appears to suit the data set, or sets, to hand. If there are several potential explanatory variables but no model has been suggested by precedence, then multiple linear regression, with the difficulties noted in Section 1.2, may be tried. There does not appear to be any useful alternative to this at present.

1.4.1   The Criterion of Least Squares

Although it would be possible to estimate the p elements of the parameter vector θ from only p data sets (that is, $n = p$) by solving p simultaneous equations, this procedure rarely provides satisfactory estimates. Instead, it is customary to use more data points (usually many more) than parameters and to estimate the parameters using the criterion of *least squares*. To understand how to compute the least-squares estimates, let us reconsider the standard regression model (1.1.3).

Other ways of writing (1.1.3) are

$$E(Y_t) = f(X_t; \theta) \qquad\qquad (1.4.3)$$

and

$$y_t = f(X_t, \boldsymbol{\theta}) \tag{1.4.4}$$

omitting the error term ϵ_t. Mathematically, $E(Y_t)$ means the *expectation* or *average value* of the response Y_t, and use of the lowercase y_t means the same thing. The different forms (1.1.3), (1.4.3), and (1.4.4) are equivalent ways of stating the same assumptions.

The term ϵ_t, appearing in (1.1.3) but omitted from (1.4.3) by use of the expectation operator or from (1.4.4) by use of the lowercase symbol, is usually assumed to be normally distributed, that is, according to a Gaussian distribution, with mean zero and constant variance σ^2 for all values of the explanatory variable X_t. One usually writes this as

$$\epsilon_t \sim N(0, \sigma^2) \tag{1.4.5}$$

(meaning that the error is assumed to be normally distributed with mean zero and variance σ^2) and refers to this stochastic assumption as an additive error assumption. For a multiplicative error assumption, one would have

$$\log Y_t = \log f(X_t; \boldsymbol{\theta}) + \epsilon_t \tag{1.4.6}$$

obtained from

$$Y_t = f(X_t; \boldsymbol{\theta})\epsilon'_t \tag{1.4.7}$$

where ϵ_t has the distribution given by (1.4.5) and $\epsilon'_t = \exp(\epsilon_t)$. Alternative ways of writing (1.4.6) are

$$E(\log Y_t) = \log y_t = \log f(X_t; \boldsymbol{\theta}) \tag{1.4.8}$$

by analogy to (1.4.3) and (1.4.4). The multiplicative assumption tends to be valid when the variability of Y_t increases with increasing values of Y_t. Figure 1.1 shows data conforming to such an assumption, where the variance σ^2 is large when Y_t is large and small when Y_t is small.

For an additive error assumption, the least-squares estimators $\hat{\boldsymbol{\theta}}$ of $\boldsymbol{\theta}$ are obtained as the values of $\boldsymbol{\theta}$, which minimize the following sum of squares:

$$S(\boldsymbol{\theta}) = \sum_{t=1}^{n} [Y_t - f(X_t; \boldsymbol{\theta})]^2 \qquad (1.4.9)$$

For a multiplicative error assumption, one minimizes

$$S(\boldsymbol{\theta}) = \sum_{t=1}^{n} [\log Y_t - \log f(X_t; \boldsymbol{\theta})]^2 \qquad (1.4.10)$$

Henceforth in this handbook, the subscript t will be dropped for simplicity and the explanatory variable written simply as X.

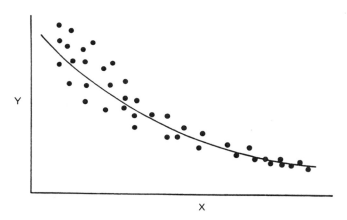

Figure 1.1 Data Conforming to a Multiplicative Error Assumption.

1.4.2   The Stochastic Term ϵ

A regression model such as (1.1.3) can be thought of as being made up of a *deterministic* part $f(X;\theta)$ and a stochastic part ϵ. Although a wide variety of assumptions about ϵ are possible, the most frequent assumption about it is that it is independently and identically normally distributed with a constant but unknown variance σ^2. The meaning of this will be described in detail in following subsections, which treat separately the three aspects, independence, identicalness, and normality.

Independence

The assumption of independent, identically distributed normal error refers to the relationship among the separate pairs of measurements (X_i, Y_i), $i = 1, 2, \ldots, n$, that is, (X_1, Y_1), (X_2, Y_2), ..., (X_n, Y_n). Independence means that the error at one value of i (say, $i = 2$) is not related to the error at some other value of i (say, $i = 7$). This should be the case when the explanatory variable is a controlled variable, as in the example of the yield of a chemical reaction as a function of the temperature of the reaction chamber, provided that the experimenter conducts each experimental trial in exactly the same fashion, allowing the experimental apparatus to return to the equilibrium or set point before going on to the next trial. This is an attempt to eliminate any carryover effects between trials. Carryover effects can occur when the same experimental units are used more than once. For example, consider a pharmaceutical trial in which human volunteers, suffering from a certain disease or ailment, agree to self-administer a series of potentially beneficial drugs. These may include a standard drug and/or a placebo, but the nature and identity of the drugs are unknown to the volunteers and the drugs are administered in some random sequence or in a predetermined sequence devised by the experimenter. If one (or more) of the drugs has a persistent effect, and if insufficient time has been allotted between the taking of the drugs for the effect of a persistent drug to wear off, the assumption of independence is violated.

A further example of correlated errors occurs in forestry inventory trials, where the total wood production by the trees on a number of plots is measured at specified intervals in order to build a model to predict the likely wood yield at some future time. Because the same plot is measured each time, the errors, that is, the departures of the measured values from values given by the model fitted over several plots, may be strongly correlated within any one plot. This follows from the fact that the characteristics of any one plot may determine that observations within that plot are all likely to lie above or below the average for all plots, which is represented by the fitted model. Problems associated with the testing of hypotheses when there are multiple measurements on the same experimental units are discussed by West et al. (1984) and Dielman (1983), and some solutions to these problems are given in West et al. (1986), but these are outside the scope of this handbook.

Identicalness

Given that the stochastic term ϵ is truly independent between pairs of measurements, it is necessary to determine whether the distribution is identical for all pairs. The question of normality (that is, whether the probability distribution for ϵ is Gaussian) is treated in the next subsection. The other aspect of identicalness is whether the moments of the distribution, particularly the variance, are constants. Consider Figs. 1.2a and 1.2b. These show data for which the deviations in a vertical direction of the data points from the model (shown by the solid line) do not appear to be a function of the magnitude of the response variable Y. However, in Figs. 1.2c and 1.2d, it is clear that the magnitude of the deviations from the model depend on the magnitude of Y (not X).

If the variance of the stochastic term increases or decreases with the magnitude of the response variable is some systematic way, the least-squares regression equations may be modified to take account of the weighting. Thus, instead of minimizing

$$S(\boldsymbol{\theta}) = \sum_{t=1}^{n} [Y_t - f(X_t; \boldsymbol{\theta})]^2$$

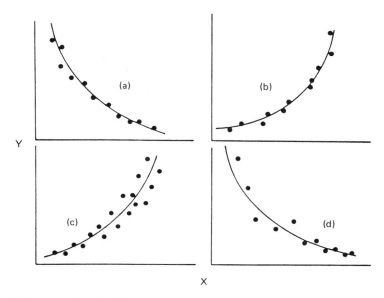

Figure 1.2   (a,b) Data with an Identical Error Distribution, and (c,d) Data Where the Magnitude of the Error Depends on Y.

the usual least-squares criterion (1.4.9), one minimizes

$$S(\boldsymbol{\theta}) = \sum_{t=1}^{n} w_t [Y_t - f(X_t; \boldsymbol{\theta})]^2 \tag{1.4.11}$$

where w_t, the weighting corresponding to the point (X_t, Y_t), is taken to be inversely proportional to the variance of the error term. Thus, each of the n data points has its own characteristic variance in the weighted least-squares regression given by (1.4.11). For example, consider data for basal area increment of trees in a regrowth eucalypt forest versus tree diameter in Fig. 1.3. Clearly, the variation in Y increases rather markedly with increasing X. West (1980) found that, for such a forest, for relationships between basal area increment and initial diameter, residual variance

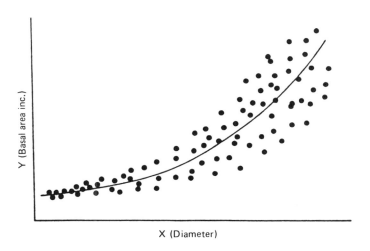

Figure 1.3   Basal Area Increment for Trees in a Regrowth Forest.

seemed to vary directly as the fourth power of the initial diameter, and weighted least-squares regressions were fitted accordingly. Weighted least-squares is discussed in textbooks such as that by Draper and Smith (1981) and will not be considered further in this handbook. It should be noted that weighting changes the estimates of the parameters and the standard errors of the estimates relative to the values obtained in the absence of weighting. However, only when there is a marked inhomogeneity of variance, such as in Fig. 1.3, will the differences between the weighted solution and the standard (unweighted) solution be considerable. For most sets of practical data, it is often very difficult to detect, with any confidence, evidence of inhomogeneity of variance.

Normality

The question of whether an error distribution is normal (Gaussian) is very difficult to resolve unless one has large samples. With large samples, the so-called *q-q* plots of the expected normal quantiles versus the residuals from the fitted model may reveal departures from the normality assumption (Wilk and Gnanade-sikan, 1968). With small samples, detection of nonnormality may be extremely difficult, especially when the number of parameters to be estimated is a sizable fraction of the sample size. Under these circumstances, the use of the least-squares criterion tends to induce "supernormality," making the residuals appear more normal than they should (Gnanadesikan, 1977). Nevertheless, the graphical checks have this benefit: they can often reveal outliers, that is, points that are, in some sense, atypical of the remaining points. Some formal significance testing of the normality assumption has been advocated (Lawless, 1982), and further research in this area must be expected. Other useful references on diagnostics are the books by Belsley et al. (1980) and Cook and Weisberg (1982).

Another approach to the examination of data in an attempt to understand its basic features is called exploratory data analysis (Tukey, 1977). Yet another procedure, which is designed to make as few assumptions about the basic features of the data as possible, is called robust data analysis, which includes robust regression. This procedure and exploratory data analysis are both discussed in a book edited by Hoaglin et al. (1983). However, these techniques are outside the scope of this handbook, which is devoted to dealing with models in which the assumption of normality of the stochastic term is a good approximation to the truth.

Summary

One may say that the standard assumption of independent, identically distributed normal error is an idealization that real data may approach in practice, without ever exactly attaining. In practice, all that is really necessary is that these assumptions be *approximately* true, as the least-squares criterion tends to be reasonably robust to minor departures from these assumptions. Only when there are major departures from the standard assumptions, such as, for example, when there is a prominent outlier, a marked inhomogeneity of variance, or a clear serial dependence between residuals, will this lead to significant errors in the estimates. Much useful information on the standard assumptions and their implications is to be found in the book by Bates and Watts (1988, Sec. 1.3.1 and Chap. 3).

2

Nonlinear Regression Modeling

2.1 LINEAR VERSUS NONLINEAR LEAST-SQUARES

Given the validity, or approximate validity, of the assumption of independent and identically distributed normal error, as discussed in Section 1.4.2, one can make certain general statements about the least-squares estimators in linear and nonlinear regression models. For a linear regression model, such as (1.1.2) or (1.1.4), the estimates of the parameters [α, β, and γ in (1.1.4) and α and β in (1.1.2)] are unbiased, are normally distributed, and have the minimum possible variance among a class of estimators known as regular estimators. This minimum possible variance is also called the minimum variance bound. These properties are generally accepted to be the most desirable properties that estimators may have. Other estimators, although they may also be unbiased, are less "efficient" in that their variances will exceed those of the least-squares estimators.   That is why the least-squares estimators are said to be minimum variance estimators. Because the least-squares estimators have a normal distribution, confidence limits and confidence regions for the estimators are readily constructed.

Nonlinear regression models differ from linear regression models in that the least-squares estimators of their parameters are not unbiased, normally distributed, minimum variance estimators. The estimators achieve this property only asymptotically, that is, as the sample sizes approach infinity.   Some nonlinear regression models have estimators that are badly biased with a highly asymmetric long-tailed nonnormal distribution, and with sample variances greatly in excess of the minimum variance bound. Generally, the smaller the sample size, the greater the extent of nonlinearity.

2.1.1   The Concept of a Close-to-Linear Nonlinear Regression Model

Nonlinear regression models differ greatly among themselves with respect to the extent to which the behavior of their least-squares estimators approximates the asymptotic properties. For example, there are some nonlinear regression models whose estimators, even when the sample sizes are relatively small, come close to being unbiased, normally distributed, minimum variance estimators.   I termed such nonlinear regression models *close-to-linear models* (Ratkowsky, 1983, Section 1.4).   Models not possessing these properties may be termed *far-from-linear*. There are nonlinear regression models that may not be close-to-linear in certain commonly used parameterizations but that, as a result of reparameterization, become close-to-linear. Some basic principles of reparameterization will be discussed in Section 2.3.   Finally, there are some nonlinear regression models that never behave in a close-to-linear fashion, even for relatively large sample sizes.   Some of these will be discussed in Section 2.5.4.

2.1.2 Convergibility

The least-squares estimates of the parameters in nonlinear regression models, unlike those in linear regression models, cannot be determined from an explicit mathematical expression. Instead, one must obtain the minimum sum of squares by some other means, such as an iterative method beginning with a set of initial parameter estimates. It is outside the scope of this handbook to discuss the various methodologies available for doing this, for which there is an enormous literature (for example, see Chambers, 1973, or Schmidt, 1982). However, the concept of ready convergibility, that is, the ease of convergence from the initial parameter estimates, and its relationship to the extent to which a model exhibits close-to-linear behavior, is well illustrated by considering only the Gauss–Newton method (for example, see Bard, 1974), which has the advantage over methods such as the Newton–Raphson method of requiring only first derivatives of the model with respect to the parameters. The convergibility of a model/data set combination depends strongly on the shape of the contours of equal residual sum of squares on the response surface drawn in parameter space. (See Ratkowsky, 1983, for an explanation of these terms.)

A linear regression model has a sum-of-squares surface in which the contours of equal residual sum of squares are elliptical (ellipses for a two-parameter model, ellipsoids or hyperellipsoids for three or more parameters). If one were to use the Gauss–Newton method to obtain the least-squares estimates for a linear model (of course, such use is unnecessary since there is an explicit formula for obtaining the estimates in that case), convergence to the least-squares estimates would be obtained in a single step, irrespective of the initial estimates, no matter how remote they may be from the least-squares estimates (Ratkowsky, 1983, Appendix 2.A). The same is true of certain methods that require second derivatives, such as the Newton–Raphson method. With nonlinear regression models, on the other hand, convergibility depends strongly on the parameterization of the model and on the initial estimates. As the extent of nonlinear behavior increases, the ease of obtaining convergence decreases since the contours of equal residual sum of squares depart more and more from the elliptical shape. Unless the initial estimates are reasonably close to the least-squares estimates, the iterative process may diverge. The question of whether convergence occurs is generally much more important than the question of what number of iterations is required to achieve convergence. Reparameterization of a basic nonlinear regression model [(for example, (1.4.2) is a reparameterization of (1.4.1)] to a close-to-linear parameterization will improve the model's convergibility. This does not necessarily mean that the close-to-linear parameterization will converge significantly more rapidly, that is, in many fewer iterations, than a far-from-linear one, but its *chances* of convergence from a reasonable starting point (that is, good initial estimates) are significantly improved.

It is not only the "almost ensured" convergence that makes one prefer a close-to-linear model to a far-from-linear one. Far-from-linear models may have more that one minimum in their residual sum-of-squares surface. Such a model may converge, if convergence occurs at all, to an optimum that is not a global optimum. Interpretation of the results becomes more difficult and suspect in that case.

Obtaining almost ensured convergence is of particular benefit when one has several sets of data for which one wishes to test for parameter invariance. That is, one may wish to examine whether a given parameter (or set of parameters) has constant values in two or more data sets. Provided the intrinsic nonlinearity (see Sec. 2.2) is adequately low, an almost exact test of the hypothesis of parameter equality is carried out using the extra-sum-of-squares principle (Draper and Smith, 1981). (See Gallant, 1987, for a comprehensive approach to the testing of hierarchical models.) To perform this test, it is necessary (1) to fit the model to each data set separately and (2) to fit the model to the combined data set, estimating common values for each parameter of interest. In this restricted estimation problem, convergence to the global optimum is extremely unlikely to be achieved unless a close-to-linear parameterization is used. In the absence of convergence, the significance test that I described in detail earlier (Ratkowsky, 1983, Chap. 7) cannot be performed.

2.1.3 Properties of Close-to-Linear Models

A close-to-linear model has many advantages over those not possessing that property. In addition to providing almost ensured convergence, the predicted values of the response variable Y in such models are virtually unbiased. The least-squares estimators of the set of parameters will have distributions closely approximating a normal distribution. Joint confidence regions for its parameters, which are almost ellipsoidal, may be constructed from the asymptotic covariance matrix (see Sec. 2.4). Similarly, confidence intervals for each parameter will be close to being exact (see Sec. 2.4). Rough comparisons of parameter values from results obtained by different workers can be made by noting the magnitudes of the parameter estimates and their asymptotic standard errors. [The correct approximate test is to use the extra-sum-of-squares principle (see Sec. 2.1.2), but the original data from other workers may not have been published or may not be readily available.] A far-from-linear model will have estimates and standard errors that will be grossly biased, so that these rough comparisons may lead to grossly incorrect inferences. The estimators of the parameters in a far-from-linear model will be nonnormally distributed, and this presents a variety of problems, especially as the degree and extent of nonnormality is generally unknown. The amount of nonnormality depends also on the sample size, so that comparisons made with samples of differing sizes are likely to contain differing degrees on nonnormality.

2.2 MEASURES OF NONLINEAR BEHAVIOR

Previously (Ratkowsky, 1983), I described a number of measures and procedures for studying the estimation behavior of nonlinear regression models. These included the curvature measures of intrinsic and parameter-effects (PE) nonlinearity of Bates and

Watts (1980), the bias measure of Box (1971), and the asymmetry measure of bias of R.K. Lowry. The latter measure has now been superseded, and it is best to use a direct measure of skewness due to Hougaard (1985). A brief description of each measure is given in Sections 2.2.1 and 2.2.2. For more details on the Bates and Watts (1980) and Box (1971) measures, see Ratkowsky (1983, Chap. 2). The use of simulation studies is described in Section 2.2.3. Computer programs for calculating those measures appear in the Appendix to Ratkowsky (1983). A subroutine for calculating the Hougaard (1985) skewness measure appears in the Appendix to this handbook (subroutine SKEW).

2.2.1   The Bates and Watts Curvature Measures

The intrinsic nonlinearity (IN) measures the curvature of the solution locus in sample space, where the locus represents all possible solutions to the estimation problem. The least-squares solution is the point on the solution locus closest to the observed response vector Y. For a linear regression model, IN is zero, since the solution locus is straight (a line, plane, or hyperplane). For a nonlinear regression model, the solution locus is curved, with IN measuring the extent of that curvature. Fortunately, as found by Bates and Watts (1980) and Ratkowsky (1983), and in many subsequent studies by this author, IN is typically small for almost all models of practical interest. This means that if a model is found to be far-from-linear, the major contribution to the nonlinearity is likely to be due to the parameterization, and a reparameterization should then be sought. As the shape of the solution locus is independent of the parameterization, the process of reparameterization does not alter IN.

The parameter-effect nonlinearity is a measure of the lack of parallelism and the inequality of spacing of parameter lines on the solution locus at the least-squares solution. [Actually, Bates and Watts (1980) used the projection of the parameter lines onto the tangent plane to the solution locus at the least-squares solution.] The PE value is a scalar quantity (that is, a single number), representing the maximum value of the effect of the parameterization, obtained from a three-dimensional array called the acceleration array. This array is a scaled version of the matrix of second derivatives with respect to the parameters. For a linear regression model, the parameters appear linearly, so that the second derivatives are all zero, resulting in a PE value of zero. For nonlinear regression models having a given value of IN, PE increases the more the model departs from a linear model. At least, this is what should happen if PE is to be a reliable indicator of the extent of nonlinearity caused by the parameterization. Unfortunately, however, PE occasionally indicates a high degree of nonlinear behavior when simulation studies indicate close-to-linear behavior (Ratkowsky, 1983, Sec. 9.5). On the other hand, a low value of PE is informative, although Cook and Witmer (1985) have found an instance in which more nonlinearity is actually present than is indicated by the PE value. Provided IN is low, it should be the goal of reparameterization to find a parameterization that makes PE as low as possible. As an approximate guide to what constitutes a low IN or PE value, one may use $1/\sqrt{F}$ or $1/\sqrt{2F}$ to test both values, where F is drawn from a table of the F-distribution, with the numerator degrees of freedom equal to the number of parameters, and the denominator degrees of freedom equal to the number of data points minus the number of parameters. More recently, Cook and Goldberg (1986) developed curvature measures for arbitrary subsets of the parameters in a nonlinear regression model. These include measures for the individual parameters themselves and thus identify which parameters are in need of reparameterization. The computational effort to obtain the subset measures is considerable, however.

Another recent development is that of profile t plots, profile traces, and profile pair sketches, due to Bates and Watts (1988). Profile t plots graphically reveal the extent of nonlinearity in individual parameter estimates, and plots of the profile traces provide useful information on how the parameters interact. Profile pair sketches provide a computationally economical method of obtaining accurate approximations to two-dimensional projections of the likelihood region. This enables contours of pairs of parameters, such as (θ_1, θ_2), (θ_1, θ_3), and (θ_2, θ_3), to be drawn.

2.2.2 Measures of Bias and Skewness

The bias in the least-squares estimator for each parameter separately, for any nonlinear regression model, is readily calculated. Cook et al. (1986) showed that the various bias approximations that have been proposed (Cox and Snell, 1968; Box, 1971; Clarke, 1980; Amari, 1982; Hougaard, 1985) are identical for the standard nonlinear regression model under consideration here. Although the bias has use as a measure of the extent to which parameter estimates may exceed or fall short of the true parameter values, it cannot be used to compare parameters in two different parameterizations because the parameterizations have different "locations" and "scales." Earlier, I employed percentage bias, the bias as a percentage of the estimate (Ratkowsky, 1983, pp. 20–23), but this quantity is not location-independent, that is, it is possible to obtain a high percentage bias simply because the estimate is near zero.

A location- (and scale-) independent measure of bias and, concomitantly, nonlinear behavior of the estimator for each parameter, was conceived by R.K. Lowry as I described earlier (Ratkowsky, 1983, Sec. 2.9). There, an asymmetry statistic ψ_i was defined as

$$\psi_i = [(\hat{\theta}_i^+ - \theta_i) + (\hat{\theta}_i^- - \theta_i)]/2 \qquad (2.2.1)$$

where $\hat{\theta}_i^+$ and $\hat{\theta}_i^-$ are, respectively, the least-squares estimators of the ith parameter θ_i obtained by adding and subtracting a vector of random normal errors with constant variance σ^2 to and from a hypothesized model. The variance of this statistic, $\mathrm{Var}(\psi_i)$, was written as an expression containing the variances of $\hat{\theta}_i^+$ and $\hat{\theta}_i^-$ and their covariance. After rearrangement, a formula for the correlation coefficient between $\hat{\theta}_i^+$ and $\hat{\theta}_i^-$ was obtained, which produces values near -1 for close-to-linear models. Because of the lack at that time of an algebraic expression for $\text{Var}(\psi_i)$, a simulation study was needed to estimate values of the correlation coefficient. The necessary algebraic expression for $\text{Var}(\psi_i)$ was subsequently derived by R. Morton (Lowry and Morton, 1983), rendering a simulation study unnecessary. Its computation requires formulas for the first and second derivatives of the model with respect to the parameters, which are also needed to calculate the IN and PE measures of Bates and Watts (1980) and the bias calculation of Box (1971). Instead of calculating the correlation coefficient between $\hat{\theta}_i^+$ and $\hat{\theta}_i^-$, Lowry and Morton (1983) used the ratio $\lambda_i = \text{Var}(\psi_i)/\text{Var}(\hat{\theta}_i)$, where $\text{Var}(\hat{\theta}_i)$ is the variance of $\hat{\theta}_i^+$ or $\hat{\theta}_i^-$, these variances being identical. The ratio λ_i is termed the asymmetry measure of nonlinearity and has values between 0 and 1. The latter measure has been subsequently developed further (Morton, 1987), but it has recently been found (Ratkowsky, unpublished results) to give incorrect assessments of the extent of nonlinear behavior for some parameters in some models. There is also no objective criterion for deciding whether λ_i is large or small and, hence, whether the extent of nonlinearity in a parameter estimator is considerable or trivial. Therefore, it is best to use a direct measure of skewness due to Hougaard (1985), which will now be described.

To calculate the Hougaard (1985) measure of skewness, we let a term like L^{ik} denote an element of

$$L = [J^T(\hat{\theta})J(\hat{\theta})]^{-1} \qquad\qquad (2.2.2)$$

where $J(\hat{\theta})$ is the $n \times p$ Jacobian matrix (see Ratkowsky, 1983, Appendix 2.A, for a detailed definition of this matrix) with typical element J_{mj}, evaluated at $\hat{\theta}$. Also, let $s^2 = \text{RSS}(\hat{\theta})/(n-p)$ be the estimate of the residual variance σ^2 based on the residual sum of squares at $\hat{\theta}$, and let $H(\hat{\theta})$ be the $n \times p \times p$ matrix of second derivatives with respect to the parameters, with typical element

H_{mkl}, evaluated at $\hat{\theta}$. Then, if W_{jkl} is a term defined by

$$W_{jkl} = \sum_{m=1}^{n} J_{mj} H_{mkl}$$

an estimate of the third moment is given by

$$E[\hat{\theta}_i - E(\hat{\theta}_i)]^3 = -(s^2)^2 \sum_{jkl} L^{ij} L^{ik} L^{il} (W_{jkl} + W_{kjl} + W_{lkj}) \quad (2.2.3)$$

with the indices j, k, and l each ranging from 1 to p. The third moment may conveniently be standardized using the appropriate element of the asymptotic covariance matrix s^2L to give

$$g_{1i} = E[\hat{\theta}_i - E(\hat{\theta}_i)]^3 / (s^2 L^{ii})^{3/2} \quad (2.2.4)$$

which provides a direct measure of the skewness of $\hat{\theta}_i$. Because the g_1 measure is a familiar one in statistics and because of the close link between the extent of nonlinear behavior of an estimator and the extent of nonnormality in the sampling distribution of the estimator, it is relatively easy to devise a rule-of-thumb for asserting whether the estimator $\hat{\theta}_i$, as assessed by g_{1i}, is close-to-linear or contains considerable nonlinearity. Thus, it is possible to say that if $g_{1i} < 0.1$, the estimator $\hat{\theta}_i$ of parameter θ_i is very close-to-linear in behavior and, if $0.1 < g_{1i} < 0.25$, the estimator is reasonably close-to-linear. For $g_{1i} \geq 0.25$, the skewness is very apparent, and $g_{1i} > 1$ indicates considerable nonlinear behavior. A listing of a FORTRAN subroutine (SKEW) for computing g_{1i} for each parameter in a model appears in the Appendix to this handbook.

2.2.3 Simulation Studies

Previously, I described simulation studies in some detail (Ratkowsky, 1983, Sec. 2.6). These studies can be extremely helpful in giving users a good idea of the behavior of the estimators without requiring them to calculate the Bates and Watts (1980) measures, the bias measure of Box (1971), or the Hougaard (1985) skewness measure. With the wide spread availability of personal computers and great advances in the speed of such computers, simulation studies can be readily performed, provided one has a program for finding the least-squares estimates of the parameters in a nonlinear regression model. To investigate the properties of a particular model/data set combination, one must first obtain the least-squares estimates and an estimate of the residual variance about the regression line. Using these estimates as though they were the true values, 500 or 1000 data sets are generated by allowing the values of the error term to change randomly (while having a normal distribution with the required variance) at each value of the explanatory variable X for each data set. The least-squares estimates of the parameters are then obtained for each of the generated data sets. Their distribution can be examined using formal tests (Ratkowsky, 1983, Sec. 2.6) or by plotting histograms of the results for each parameter separately. Histograms readily reveal nonnormal behavior because an asymmetric histogram is sufficient evidence of nonnormality. Thus, simulation studies can provide an alternative to calculating the various measures of nonlinear behavior described in Sections 2.2.1 and 2.2.2 although the computation time for doing a simulation study of 500 trials is considerably greater than the time for calculating those measures.

2.3 SOME BASIC PRINCIPLES OF REPARAMETERIZATION

The models (1.4.1) and (1.4.2) are "reparameterizations" of each other; that is, the parameters of one of them may be expressed as a function *only* of the parameters of the other model, without the expression containing the explanatory variables, the response variables, or the error term. It is a consequence of least-squares estimation that the mathematical function relating the estimators in the two models is the same mathematical function as that relating the parameters. Using a circumflex over the parameter to indicate the least-squares estimator of that parameter, it follows that $\hat{\theta}_1 = 1/\hat{\alpha}$ and $\hat{\theta}_2 = \hat{\beta}/\hat{\alpha}$, and the fitted or predicted values from the two parameterizations are identical. That is why models like (1.4.1) and (1.4.2) were referred to in Section 1.4 as the same *basic* model. Although different parameterizations of the same basic model produce the same goodness-of-fit and the same fitted values, etc., they may differ greatly in their estimation behavior.

In Section 2.2.1, the concept of a close-to-linear nonlinear regression model was discussed, and reasons were given why such models should be preferred to models not having that property. Sometimes it is only a single parameter in an nonlinear regression model that prevents that model from exhibiting close-to-linear behavior. If a suitable parameterization can be found for that parameter, it is possible to change a model's far-from-linear behavior to close-to-linear behavior.

Although reparameterization is desirable to improve the behavior of a basic nonlinear regression model, there were previously few guidelines to assist a modeler in choosing ways of replacing offending parameters by ones with better properties. One way that was suggested (see Ratkowsky, 1983, p.39) was to perform a simulation study and look at histograms of the estimates of each parameter separately. Those parameters having estimates whose distributions are close to that of a normal distribution are already close-to-linear and do not require reparameterization. Those whose estimates have long-tailed distributions are candidates for reparameterization. If the distribution has a long right-hand tail, resembling the logarithm of a normally distributed random variable, the parameter could then be replaced by the exponential of, or a positive integer power of, a new parameter. However, if the distribution of the estimates has a long left-hand tail, suggestive of the exponential of a normally distributed random variable, the parameter might then be replaced by the logarithm of, or a positive fractional power of, a new parameter. Aside from these vague guidelines, there was no theory to tell the user whether these attempts at reparameterization would dramatically reduce the extent of nonlinear behavior. The next section shows how this situation has changed with the realization that expected-value parameters would invariably lead to parameterizations with good estimation properties.

2.3.1 Expected-Value Parameters

A class of parameters that exhibits close-to-linear behavior is the class of *expected-value* parameters. These parameters are of the type suggested by Ross (1975) and correspond to the fitted (predicted) values of the response variable Y. Since the bias in predicted values depends only on the intrinsic nonlinearity (Ratkowsky, 1983, Sec. 9.2) and since the intrinsic nonlinearity is typically small for almost all models of interest (Sec. 2.2.1), this bias is small. At the same time, the extent of nonnormality and excess variance of the least-squares estimates of an expected-value parameter is also small. The only restriction on expected-value parameters is that they should fall within the observed range of the data and not correspond to asymptotes or extrapolations outside the data range. Expected-value parameters outside the range of the observed data are less efficacious.

Finding Expected-Value Parameters

To find expected-value parameters, one chooses p values of the explanatory variable X, p being the number of parameters. The new parameters are the expected values, denoted $y_1, y_2, \ldots, y_p$, after the original parameters are eliminated. The procedure is illustrated using the two-parameter model

$$y = \alpha X^{\beta} \tag{2.3.1}$$

The first step is to choose values X_1 and X_2 of the explanatory variable X, making sure that they fall within the observed range of the variability of X (they *may* be the minimum and maximum observed values of X in the data set). We get the expected values

y_1 and y_2 from

$$y_1 = \alpha X_1^\beta \qquad\qquad (2.3.2)$$

and

$$y_2 = \alpha X_2^\beta \qquad\qquad (2.3.3)$$

As (2.3.2) and (2.3.3) are two equations in two unknowns, α and β, one can solve for those parameters to obtain

$$\alpha = y_1 / X_1^{\log(y_1/y_2)/\log(X_1/X_2)} \qquad\qquad (2.3.4)$$

and

$$\beta = \log(y_1/y_2)/\log(X_1/X_2) \qquad\qquad (2.3.5)$$

which, substituting back into the original equation (2.3.1), yields

$$y = y_1 (X/X_1)^{\log(y_1/y_2)/\log(X_1/X_2)} \qquad\qquad (2.3.6)$$

which may be rewritten as

$$y = y_1 (y_1/y_2)^{\log(X/X_1)/\log(X_1/X_2)} \qquad\qquad (2.3.7)$$

The reader can easily verify that when $X = X_1$, $y = y_1$, and that when $X = X_2$, $y = y_2$. The procedure has eliminated the original parameters α and β in favor of two new parameters y_1 and y_2 which, by virtue of being expected-value parameters, will have close-to-linear estimation behavior.

Equation (2.3.7) is not a particularly pretty equation, and it contains one parameter (y_1) twice. This is a feature of parameterizations containing expected-value parameters: the expressions are more cumbersome in appearance than the original expressions, and the expected-value parameters may appear more than once. As compensation for this loss of aesthetics, however, parameterizations with expected-value parameters offer three advantages. First, one obtains rapid convergence to the least-squares estimates using the new parameterization, since the new model is close-to-linear (Sec. 2.1.2). A second benefit is that initial parameter estimates are very easy to obtain. One may simply draw a free-hand curve through the graph of the data for Y versus X and read off approximate fitted values y corresponding to the chosen values X_1, X_2, etc., for X. Third, the expected-value parameters are more suitable for inference than the original parameters since their least-squares estimates are close to being unbiased, normally distributed, minimum variance estimators.

Limitations of Expected-Value Parameters

Lest the reader think that the parameters of all basic models can be replaced by expected-value parameters to provide an analytical expression containing only expected-value parameters, we consider the following model

$$y = \alpha(1 - \beta^X) \tag{2.3.8}$$

As before, one chooses two values of X, X_1, and X_2 (say), to obtain

$$y_1 = \alpha(1 - \beta^{X_1}) \tag{2.3.9}$$
$$y_2 = \alpha(1 - \beta^{X_2}) \tag{2.3.10}$$

Although α can be eliminated from (2.3.9) and (2.3.10) to yield

$$y = (1 - \beta^X)y_2/(1 - \beta^{X_2}) \tag{2.3.11}$$

β cannot be eliminated. Thus, algebraic limitations prevent the universal application of expected-value parameters. However, as it is often the case that one or two parameters only may be responsible for the far-from-linear behavior of a nonlinear regression model, it is seldom necessary to replace all the parameters of a model by expected-value parameters. For example, consider the Weibull-type model in the parameterization

$$y = \alpha - \beta \exp(-\gamma X^\delta) \tag{2.3.12}$$

It was found (Ratkowsky, 1983, Sec. 4.4.5) that γ was the only parameter of the four parameters in (2.3.12) that gave rise to a substantially nonnormally distributed estimator. It was shown there that the reparameterization obtained by replacing γ by its exponential to yield

$$y = \alpha - \beta \exp[-\exp(-\gamma)X^\delta] \tag{2.3.13}$$

had better properties than γ in (2.3.12). However, an alternative is to replace γ by an expected-value parameter. Hence, we can choose a value X_1, preferably somewhere near the mean of the observed X-values in the data set, to obtain, from (2.3.12),

$$y_1 = \alpha - \beta \exp(-\gamma X_1^\delta) \tag{2.3.14}$$

Solving this equation for γ, and substituting it into (2.3.12), yields

$$y = \alpha - \beta[(\alpha - y_1)/\beta]^{(X/X_1)^\delta} \tag{2.3.15}$$

which is not a cumbersome equation and which has excellent statistical properties in estimation.

2.3.2   Parameters-in-Denominator Principle for Catalytic Chemical Reactions

Catalytic chemical reactions are often of a form in which there are one or more constants in the numerator of the expression, for example,

$$r = k K_A K_B p_A p_B / (1 + K_A p_A + K_B p_B)^2 \qquad (2.3.16)$$

where k is a rate constant, K_A and K_B are equilibrium constants, and p_A and p_B are the explanatory variables, in this case partial pressures of the reactants A and B, which are assumed to be either fixed or measured accurately. Such models can be reparameterized to produce a close-to-linear model by clearing off constants in the numerator of the expression (Ratkowsky, 1985). Hence, the preceding expression becomes

$$r = p_A p_B / (\phi_1 + \phi_2 p_A + \phi_3 p_B)^2 \qquad (2.3.17)$$

where the new parameters ϕ_1, ϕ_2, and ϕ_3 are generally close-to-linear in behavior. Similar considerations apply to enzyme-catalyzed biochemical reactions. For example, the standard Michaelis–Menten model

$$v = V_{\max}[S] / \{[S] + K_m\} \qquad (2.3.18)$$

where v is the reaction velocity and $[S]$ is the substrate concentration, could be better written as

$$v = [S] / \{\theta_1[S] + \theta_2\} \qquad (2.3.19)$$

where the new parameters θ_1 and θ_2 are close-to-linear in behavior (Ratkowsky, 1986a). The same parameters-in-denominator principle applies to more complicated biochemical reactions.

2.4   CONFIDENCE REGIONS AND CONFIDENCE INTERVALS

Whatever nonlinear regression model scientists choose ultimately to use, they almost universally desire some measure of the precision with which the parameters have been estimated. Statistical techniques for obtaining these measures include the use of confidence regions and confidence intervals. Regions are used when one wishes to make a joint statement about the set of parameter values, whereas intervals apply to individual parameters separately.

2.4.1   Confidence Regions

A confidence region is a region in p-dimensional space that attempts to "cover" the true but unknown parameter vector θ with the nominal probability. One of the methods is exact, which means that the coverage probability equals the nominal probability. The other methods are approximate, meaning that the coverage and nominal probabilities do not coincide. Three methods will be described in the following subsections.

Exact Confidence Regions

For a p-parameter ($p \geq 1$) nonlinear regression model, there is an exact method of obtaining a confidence region, referred to variously as the lack-of-fit method (Gallant, 1987; Donaldson and Schnabel, 1987) or the score statistic method (Hamilton, 1986). The method dates back at least to Williams (1962), Halperin (1963) and Hartley (1964). Determination of the exact region is computationally tedious and involves finding the values of θ^*, which makes the ratio

$$R = \frac{[\mathbf{Y} - \mathbf{f}(\theta^*)]^T P(\theta^*)[\mathbf{Y} - \mathbf{f}(\theta^*)]/p}{[\mathbf{Y} - \mathbf{f}(\theta^*)]^T [I - P(\theta^*)][\mathbf{Y} - \mathbf{f}(\theta^*)]/(n-p)} \qquad (2.4.1)$$

equal exactly to $F(p, n-p; \alpha)$, the tabulated value of an F-distribution with numerator degrees of freedom p, denominator degrees of freedom $n-p$, and a specified significance level α. The "projection" matrix $P(\theta^*)$, which is $n \times n$, is given by

$$P(\theta^*) = J(\theta^*)[J^T(\theta^*)J(\theta^*)]^{-1}J^T(\theta^*) \qquad (2.4.2)$$

where $J(\theta^*)$ is the $n \times p$ Jacobian matrix (see Ratkowsky, 1983, Appendix 2.A, for a detailed definition of this matrix). The matrix I is an $n \times n$ "identity" matrix, that is, a matrix with 1s on the diagonal and 0s elsewhere. The $n \times 1$ column vector $[Y - f(\theta^*)]$ represents the residuals between the observed values of the response variable Y and the fitted values $f(\theta^*)$ using the parameter values θ^*. The boundary of the exact confidence region represents all values of θ^* that result in a value R having exactly the tabulated F-distribution value $F(p, n-p; \alpha)$.

In addition to its computational difficulty, the exact confidence regions may exhibit undesirable characteristics. For example, the region may consist of disjoint subregions.

Likelihood Confidence Regions

The likelihood, or likelihood ratio, confidence region is based on finding θ^* such that

$$L = \frac{[\text{RSS}(\theta^*) - \text{RSS}(\hat{\theta})]/p}{\text{RSS}(\hat{\theta})/(n-p)} = F(p, n-p; \alpha) \qquad (2.4.3)$$

Here, $\text{RSS}(\hat{\theta})$ is the residual sum of squares corresponding to the least-squares estimate $\hat{\theta}$ and $\text{RSS}(\theta^*)$ is the residual sum of squares corresponding to θ^*. Unlike the exact method described under "Exact Confidence Regions," Section 2.4.1, likelihood regions are approximate. Likelihood regions are generally somewhat easier to calculate than exact regions but are still computationally tedious compared with the linearization method described in the following subsection.

Linearization Confidence Regions

The most commonly used method of finding confidence regions is that based on linearization at the least-squares estimate $\hat{\theta}$. This involves finding values of θ that satisfy

$$(\theta - \hat{\theta})^T \hat{V}^{-1}(\theta - \hat{\theta}) = pF(p, n - p; \alpha) \qquad (2.4.4)$$

The estimated variance-covariance matrix $\hat{V}$ is customarily approximated by

$$\hat{V} = s^2[J^T(\hat{\theta})J(\hat{\theta})]^{-1} \qquad (2.4.5)$$

where $J(\hat{\theta})$ is the $n \times p$ Jacobian matrix evaluated at $\hat{\theta}$ (see Ratkowsky, 1983, Appendix 2.A) and where $s^2 = \text{RSS}(\hat{\theta})/(n - p)$, the estimate of the residual variance based on the residual sum of squares at $\hat{\theta}$. Other estimators have been suggested for the variance-covariance matrix $\hat{V}$, but Donaldson and Schnabel (1987) found that the definition given by (2.4.5) was preferable to other variants.

Unlike the exact and likelihood regions, which may have irregular shapes, the linearization region is always "ellipsoidal." The extent to which such a region approximates the exact region depends on the extent of nonlinearity in the model. If the parameter-effects nonlinearity is considerable, there may be big discrepancies between the linearization region and the exact or likelihood regions (see Cook and Goldberg, 1986, for some graphical representations of linearization regions compared with likelihood regions). The more close-to-linear the model, the more closely will its linearization region coincide with its likelihood and exact regions.

2.4.2 Confidence Intervals

It is generally true that users find confidence intervals for individual parameters more suitable for summarization of the results of experimentation than confidence regions for a set of parameters. Their major disadvantage is that they may individually cover the elements of the true but unknown parameter vector $\boldsymbol{\theta}$, but collectively the joint confidence region may not contain $\boldsymbol{\theta}$.

Lack-of-Fit Confidence Intervals

Unlike the corresponding lack-of-fit confidence regions (see under "Exact Confidence Regions," Sec. 2.4.1), which are always exact, lack-of-fit confidence intervals are exact only under certain circumstances. Exactness of a confidence interval for a given parameter will prevail only when the regression model is linear in the remaining parameters when the given parameter is excluded. Computationally, the lack-of-fit method is more complicated than the likelihood method, and it is also structurally more undesirable as it may result in unconnected intervals (Gallant, 1976), just as lack-of-fit confidence regions may also be disjoint. The formula for obtaining lack-of-fit confidence intervals is not presented here, but appears in Halperin (1963), and in Donaldson and Schnabel (1987), for example.

To illustrate the computations, we present results obtained using a simple data set that served for illustrative purposes in Ratkowsky (1983), p. 3. The model considered was

$$Y_t = X_t^\theta + \epsilon_t \tag{2.4.6}$$

There were two data points ($n = 2$), namely,

$$X^T = (2,3), \qquad Y^T = (2.5, 10)$$

Results are presented in Table 2.1 for various significance levels α. Since $p = 1$, the confidence intervals are also confidence regions and are therefore exact. They were computed using (2.4.1).

Table 2.1 Lack-of-Fit
Confidence Intervals for θ
in Illustrative Problem

α	Lower limit	Upper limit
0.05	$-\infty$	3.604
0.10	$-\infty$	2.836
0.25	1.482	2.386
0.50	1.876	2.202

Likelihood Confidence Intervals

The likelihood or likelihood ratio confidence interval for the jth
parameter θ_j is obtained by finding the values of $\boldsymbol{\theta}^*$ such that

$$L = \frac{\text{RSS}(\boldsymbol{\theta}^*) - \text{RSS}(\hat{\boldsymbol{\theta}})}{\text{RSS}(\hat{\boldsymbol{\theta}})/(n-p)} = F(1, n-p; \alpha) \qquad (2.4.7)$$

As for the case of the likelihood confidence region (see subsection
of Sec. 2.4.1), $\text{RSS}(\hat{\boldsymbol{\theta}})$ is the residual sum of squares corresponding
to the least-squares estimate $\hat{\boldsymbol{\theta}}$. Here, however, $\text{RSS}(\boldsymbol{\theta}^*)$ is the
residual sum of squares corresponding to least-squares fitting of
the parameters $\theta_1, \ldots, \theta_{j-1}, \theta_{j+1}, \ldots, \theta_p$, given that θ_j is held
constant. The most extreme values of θ_j that result in a solution to
(2.4.7) are the required limits of the confidence interval. Applying
this procedure to the illustrative problem results in the values
tabulated in Table 2.2.

Linearization Confidence Intervals

The linearization method is the most common method of finding
confidence intervals and is the only one that has been used
routinely in computer software packages for fitting nonlinear
regression models. For the jth parameter θ_j, the limits to the

Table 2.2 Likelihood
Confidence Intervals for θ
in Illustrative Problem

α	Lower limit	Upper limit
0.05	$-\infty$	3.123
0.10	$-\infty$	2.733
0.25	1.559	2.374
0.50	1.880	2.200

confidence interval are simply

$$\theta_j^* = \hat{\theta}_j \pm (V_{jj}F)^{1/2}$$

where V_{jj} is the element in the jth row and jth column of the estimated variance-covariance matrix $\hat{V}$ and $F = F(1, n-p; \alpha)$. The above can also be written as

$$\theta_j^* = \hat{\theta}_j \pm t(V_{jj})^{1/2} \tag{2.4.8}$$

where t is the tabulated value for the t-distribution with $n - p$ degrees of freedom and significance level α.

The simplicity of (2.4.8) is attractive, but readers should remember that unless the model is close-to-linear, linearization limits may be grossly in error. For the illustrative data set, the limits obtained by linearization are given in Table 2.3. Comparing these with the lack-of-fit limits in Table 2.1 or the likelihood limits in Table 2.2 reveals large discrepancies at $\alpha = 0.05$ and $\alpha = 0.10$, with the linearization method yielding finite lower limits at these significance levels in contrast to the other methods.

Table 2.3   Linearization Limits
for θ in Illustrative Problem

α	Lower limit	Upper limit
0.05	0.053	4.055
0.10	1.059	3.048
0.25	1.673	2.434
0.50	1.896	2.211

2.5 FALLACIES, MISCONCEPTIONS, AND TRAPS

In my earlier book (Ratkowsky, 1983), I dealt with a few commonly
held fallacies in nonlinear regression modeling. In addition to
reiterating two of the fallacies in Sections 2.5.1 and 2.5.2, this
section deals with the misuse of R^2 (2.5.3) and the trap of seeking
overgenerality (2.5.4).

2.5.1 Parameter Correlation

One of the most persistent fallacies among users of nonlinear
regression models is the belief that high parameter correlation
may cause difficulties in obtaining convergence to the least-squares
estimates. However, it should be clear from the material presented
in Section 2.1.2 that convergence is related only to the shape
of the residual sums of squares (RSS) surface. The closer the
contours of equal RSS are to being ellipsoidal, the faster the
convergence. It is for that reason that a close-to-linear model has
rapid convergence. Convergence is unrelated to the orientation of
those ellipses or hyperellipses with respect to the parameter axes,
which is the factor that determines the parameter correlations.
It is true, however, that high parameter correlation is sometimes
indicative of overparameterization, which may cause convergence
problems (see Sec. 2.5.4). It would indeed be desirable to find
nonlinear regression models that had not only good estimation
behavior but, *at the same time*, low parameter correlations. In

general, though, that remains an unfulfilled wish although, with expected-value parameters (Sec. 2.3.1), one can choose the values of the explanatory variable in such a way as to make the parameter correlations low. This is a trial-and-error procedure, however.

2.5.2 Linear-Appearing and Nonlinear-Appearing Parameters

A further fallacy concerns the role of linear-appearing and nonlin-ear-appearing parameters. Whether a parameter appears linearly in a nonlinear regression model bears no relationship to its estimation behavior. Consider a model such as

$$y = \alpha(X - \beta)^\gamma \tag{2.5.1}$$

It might be thought that, as α appears linearly, it ought to exhibit close-to-linear behavior in estimation. If fact, α tends to be the worst-behaving parameter in (2.5.1). Fortunately, such parameters are readily replaced by expected-value parameters. Models such as (2.5.1) are special cases of the general form

$$y = \alpha f(X; \beta, \gamma, \ldots) \tag{2.5.2}$$

where $f(X; \beta, \gamma, \ldots)$ indicates a function of the explanatory variable X and parameters β, γ, etc. If y_1 is the expected value of Y at $X = X_1$, then α can be eliminated from (2.5.2) to yield

$$y = y_1 f(X; \beta, \gamma, \ldots)/f(X_1; \beta, \gamma, \ldots) \tag{2.5.3}$$

Similarly, α can appear linearly in the form

$$y = \alpha + f(X; \beta, \gamma, \ldots) \tag{2.5.4}$$

which leads to its replacement by the expected-value parameter y_1 (corresponding to $X = X_1$) to give

$$y = y_1 - f(X_1; \beta, \gamma, \ldots) + f(X; \beta, \gamma, \ldots) \qquad (2.5.5)$$

2.5.3 The Misuse of R^2

A further commonly held fallacy is the belief that R^2, the "proportion of explained variation," is of use in deciding whether a nonlinear regression model provides a good fit to the data. It is only when one has a linear model with a constant term in the model that R^2 genuinely represents the proportion of variation explained by the model. That is, the model must be able to be written in the form (2.5.4), where α is the constant term but where $f(X; \beta, \gamma, \ldots)$ is linear in the parameters β, γ, ... If α is not present, but $f(X; \beta, \gamma, \ldots)$ is linear, we have a linear "regression through the origin" situation, and R^2 must be interpreted differently. Other objections to R^2 have been raised (Draper, 1984; Healy, 1984; Helland, 1987). Irrespective of whether there is a constant term in the model, R^2 does not have any obvious meaning for a nonlinear regression model.

Therefore, the reader may ask how a modeler decides whether the nonlinear regression model provides a good fit to the data. Once having decided that there is no evidence of lack of independence, identicalness, and normality of the stochastic term (see Sec. 1.4.2) for the data set/model combination in question, one can only look at the magnitude of the residual variance and decide whether it is sufficiently small (or use the methods recommended by Draper, 1984, or Healy, 1984). Decisions about whether one model is better than another are seldom meaningful when there is only a single data set. Given several data sets, however, it is usually possible to decide whether one model is preferable to another or whether two or several models fit the data equally well. Examining the residuals after fitting the models by looking for runs of like-signed residuals or plotting expected normal quantiles against the residuals (Wilk and Gnanadesikan, 1968) may assist the modeler in deciding which of several competing models is the most appropriate. R^2 has no role to play in such evaluations and need never be calculated.

2.5.4 The Trap of Overgenerality

Far more serious and a far greater threat to valid nonlinear regression modeling than the fallacies discussed in Sections 2.5.1 and 2.5.2 is the belief that a complicated model, or a generalization of a commonly used model, is superior to a simple model. It is difficult to understand how such an attitude has come to be promulgated and accepted by many modelers, as it is contrary to philosophical attitudes developed centuries ago. The most celebrated fundamental tenet of scientific endeavor was enunciated in the fourteenth century by William of Ockham (also spelled "Occam"), an influential Catholic philosopher who was eventually excommunicated for his thinking, which contradicted the teachings of the Church. His famous dictum, now known as Occam's Razor and usually quoted as "Plurality should not be assumed without necessity," exhorts one to believe that simplicity is to be preferred to complexity, that parsimony is more likely to reflect reality that its opposite. This should serve as the basic principle for scientists engaged in nonlinear regression modeling as well as in other scientific areas. It is well recognized that, other things being equal, the greater the number of parameters, the greater the extent of nonlinear behavior. That mirrors what Bellman (1957) called the "curse of dimensionality" in another context, but it applies to nonlinear regression modeling as well.

To understand this point, consider a model with good statistical properties in estimation, such as the reciprocal model of Shinozaki and Kira (1956) for yield-density studies,

$$y = 1/(\alpha + \beta X) \tag{2.5.6}$$

One would expect greater nonlinearity when the model is extended to include additional terms. Thus, the Holliday (1960) model,

$$y = 1/(\alpha + \beta X + \gamma X^2) \qquad (2.5.7)$$

does have more nonlinearity than the Shinozaki–Kira reciprocal model but only slightly more nonlinearity (Ratkowsky, 1983, Sec. 3.2). If, however, the extra parameter is added as an exponent, as in the Bleasdale and Nelder (1960) model,

$$y = 1/(\alpha + \beta X)^\theta \qquad (2.5.8)$$

the nonlinearity increases markedly (Ratkowsky, 1983, Sec. 3.2). There is no way of predicting whether the addition of extra terms will increase the nonlinearity slightly or markedly. All one can say is that it will increase it and that the smaller the number of parameters, the smaller the nonlinearity is likely to be.

Another example of the differing effects of adding an extra parameter to a basic model with good estimation properties involves the logistic model

$$y = \alpha/[1 + \exp(\beta - \gamma X)] \qquad (2.5.9)$$

Adding a constant term to this model to produce the four-parameter logistic model widely used in bioassay work,

$$y = \delta + \alpha/[1 + \exp(\beta - \gamma X)] \qquad (2.5.10)$$

scarcely increases the nonlinearity (Ratkowsky and Reedy, 1986). However, should the parameter be added as an exponent, as in the Richards (1959) model,

$$y = \alpha/[1 + \exp(\beta - \gamma X)]^\delta \qquad (2.5.11)$$

the nonlinearity increases dramatically (Ratkowsky, 1983, Chap. 4).

The Richards model is a particularly unfortunate model because not only is its parameter-effects (PE) nonlinearity high but so too is its intrinsic (IN) nonlinearity (see Sec. 2.2.1 for a discussion of these measures). Because IN, which measures the curvature of the solution locus, is high and because reparameterization does not alter the shape of the solution locus, thereby leaving IN unchanged, there is no way to improve this property of the model by reparameterization. In general, estimates derived from this model tend to be severely biased for at least one, and often three, of its parameters. Some authors, such as Causton and Venus (1981), have placed great value on the Richards model as a model for plant growth. However, even a cursory examination of the parameter estimates and their variances given in the Appendix, pp. 262–275, of Causton and Venus (1981), show unsatisfactorily large variances for one of the parameters [β in the notation of (2.5.11)], the variance sometimes exceeding the estimate by an order of magnitude. Causton and Venus carried out simulation studies to examine the properties of the estimators (pp. 109–113) and, although they obtained satisfactory results for sunflower harvests, very high biases, skewness, and kurtosis were obtained for wheat harvests. The latter result is fairly typical when the Richards model is used in practice.

Visual examination of the graphs of some of the data sets shown in Causton and Venus (1981) reveals that often only the lower or upper portion of a sigmoidal curve, rather than the full curve, is present. Modelers should realize that it is ludicrous to attempt to fit a model that graphically describes a certain shape to data that do not conform to that shape. Although this should be obvious, there are numerous examples in which modelers have tried to fit a complicated model to data where a simpler one of fewer parameters would have sufficed. The attempt to fit a complicated model where a simple one suffices often fails, as overparameterization (that is, using a model with too many parameters) leads to convergence difficulties. Such overparameterized models often have multiple optima, so that convergence, if it occurs at all, may occur to the wrong minimum. The estimates, if they can be obtained, are usually severely biased and nonnormally distributed. Hence, such models should be abandoned in favor of simpler models with good statistical properties capable of fitting the data to a similar degree of accuracy, as judged by the residual variance (*not* the residual sum of squares, which will always tend to decrease as extra parameters are added). The residual variance takes account of the number of parameters and may increase as extra parameters are added if the reduction in the residual sum of squares is not sufficient to compensate for the reduction in the number of residual degrees of freedom.

Many other examples of the abuse of Occam's Razor may be cited, cases in which complicated models have been advocated where simpler ones would do. If there were no alternatives to using the complicated models that have undesirable estimation properties, this would be a very serious drawback in modeling. However, it is fortunate that alternative models exist, which have good statistical properties, as substitutes for the models with poor estimation properties. Instead of (2.5.8), one can use (2.5.7) with a great deal of confidence that that model will fit yield-density data very well, in addition to being very close-to-linear in behavior (see Ratkowsky, 1983, Chap. 3). As a substitute for the Richards model (2.5.11), one may use the Weibull-type model or the four-parameter Gompertz model, which will be presented in Section 5.3. It is a rare instance in which no adequately fitting close-to-linear model exists as a substitute for a badly behaving model.

3

An Illustrative Example of Regression Modeling

3.1 INTRODUCTION

This chapter attempts to illustrate the philosophy and principles described in the previous two chapters. The reader will see that there are various problems associated with regression analysis, which require an illustrative example to highlight them and to bring them into focus. These problems could not be adequately covered in the preceding chapters but come to the fore with the aid of an illustrative example. Material covered in this chapter includes: (1) recognition of the importance of distinguishing between the explanatory and response variables, (2) the importance of correct specification of the stochastic term, (3) the equivalence between different models as a result of a transformation of the explanatory variable, (4) the importance of parsimony in the choice of a suitable model, and (5) the resolution of the conflict between statistical versus practical considerations in the choice of a regression model.

3.2 FOOD ISOTHERMS, A CLASS OF LINEAR AND NONLINEAR REGRESSION MODELS

As an illustration of the problems a modeler might encounter in trying to select an appropriate model for practical use, we consider the class of isotherms that relate the moisture content of a food product, often on a percentage dry basis (but sometimes a percentage wet basis is used), to the water activity a_w, the ratio of the water vapor pressure in the food to the vapor pressure of pure water at the same temperature. Such isotherms are important for determining the appropriate packaging, storage conditions, and shelf life of foods. Some typical food isotherms are shown in Figure 3.1. Hundreds of isotherms, showing a wide variation

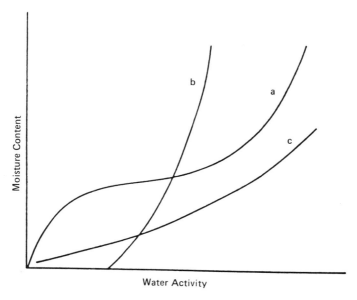

Figure 3.1   Some typical food isotherms.

in response shape, are given in a handbook of food isotherms by Iglesias and Chirife (1982). Some response curves, such as Figure 3.1a, have a sigmoidal shape, but many others do not possess an inflection point (Figs. 3.1b and 3.1c). Because of the wide variety of observed responses, it is not surprising that a very large number of mathematical models have been proposed to model the relationship between moisture content and water activity. We will confine attention in this example to just a handful of these models. They will include, however, some of the models that have been found to be very useful in practice. The models considered are to be found in the handbook by Iglesias and Chirife (1982) and in articles by Chirife and Iglesias (1978) and van der Berg and Bruin (1981).

In all the models given below, the water activity, commonly designated a_w, is denoted X here to emphasize that it is the explanatory variable. The moisture content is denoted Y, as it is customary to wish to predict moisture content at given values of water activity. In the 13 models below, the symbols a, A, b, B, C, C_1, C_2, D, E, k, n, Y_M are constants (parameters) to be estimated from a set or sets of data on water activity X versus moisture content Y. In all cases, "log" denotes the natural (Naperian) logarithm, although it generally doesn't matter which base logarithms is used.

Freundlich (1926) model:

$$Y = CX^{1/n} \tag{3.2.1}$$

BET (Brunauer et al., 1938) model:

$$X/[(1-X)Y] = 1/Y_M C + X(C-1)Y_M C \tag{3.2.2}$$

Bradley (1936) model:

$$\log 1/X = C_1 C_2^Y \tag{3.2.3}$$

Chen (1971) model:

$$X = \exp[k + a\exp(bY)] \qquad (3.2.4)$$

Halsey (1948) model:

$$X = \exp[-A/Y^B] \qquad (3.2.5)$$

Henderson (1952) model:

$$1 - X = \exp[-AY^B] \qquad (3.2.6)$$

Oswin (1946) model:

$$Y = A[X/(1-X)]^B \qquad (3.2.7)$$

Iglesias and Chirife (1981) model:

$$Y = A[X/(1-X)] + B \qquad (3.2.8)$$

Smith (1947) model:

$$Y = A - B\log(1-X) \qquad (3.2.9)$$

Kühn (1967) model:

$$Y = A/\log X + B \qquad (3.2.10)$$

GAB model (named in honor of Guggenheim, Anderson, and de Boer)

$$Y = Y_M CkX/[(1-kX)(1-kX+CkX)] \qquad (3.2.11)$$

Hailwood and Horrobin (1946) model:

$$X/Y = A + BX - CX^2 \tag{3.2.12}$$

D'Arcy and Watt (1970) model:

$$Y = AX/(1 + BX) + CX + DX/(1 - EX) \tag{3.2.13}$$

3.3   GENERAL OBSERVATIONS ABOUT THE FOOD ISOTHERM MODELS

A few general remarks can be made about the sorption isotherms of Section 3.2. The Freundlich model (3.2.1) can fit only convex/concave curves, that is, curves not having maxima, minima, or inflection points. Generally, that model is only capable of fitting data at very low water activities ($a_w < 0.15$), although there are several sets of data depicted in Iglesias and Chirife (1982) that lack an inflection point throughout the entire range of the isotherm. For water activities up to about $a_w = 0.40$, the BET model (3.2.2) seems capable of fitting most sigmoidal sorption isotherms, although there are exceptions. An alternative to the BET model is the GAB model (3.2.11), or the Hailwood and Horrobin model (3.2.12), which is nothing more than a rearrangement and reparameterization of (3.2.11). The extra parameter in these models makes the equation capable of fitting sorption data up to $a_w = 0.94$. However, the wide range of shapes that food isotherms are capable of assuming should caution us against having too high expectations that any single model will serve for all data sets over the whole range of a_w values. A recent review (Kinsella and Fox, 1986) concludes that none of the multitudinous proposed models provides accurate predictions of moisture sorption data on food materials over the complete range of a_w, reflecting the fact that sorption is a complex phenomenon involving a large number of mechanisms, which may change, for a given material, with the moisture content Y.

The form of the BET model as expressed by (3.2.2) illustrates the first basic problem confronting a modeler. Although it is usual to want to predict moisture content Y given values of the water activity X, (3.2.2) does not have Y explicitly expressed as a function of X and the parameters, to conform with the standard expression of a regression model (see Sec. 1.1). To achieve such an expression, it is necessary to rearrange (3.2.2) to give

$$Y = Y_M CX / \{[1 + X(C-1)](1-X)\} \tag{3.3.1}$$

However, (3.3.1), in common with (3.2.1) to (3.2.13), is incomplete because it lacks a stochastic term. If the variance about Y doesn't alter as X assumes different values (for example, the data depicted in Fig. 3.2 appear to satisfy this assumption), it is appropriate to add the stochastic term ϵ to the right-hand side of (3.3.1). If, however, a multiplicative error assumption is more appropriate, one would take the logarithms of both sides of (3.3.1) before adding the error term, to obtain

$$\log Y = \log\{Y_M CX / [[1 + X(C-1)](1-X)]\} + \epsilon \tag{3.3.2}$$

Whether this is more appropriate than the additive error assumption or whether some other assumption is more justifiable can only be decided by examining typical data sets for the particular problem under consideration.

Several other models among (3.2.1)–(3.2.13) do not have Y explicitly expressed as a function of X and the parameters. Consider the Halsey model (3.2.5). This may be rewritten as

$$Y = C / [\log(1/X)]^D \tag{3.3.3}$$

and, similarly, the Henderson model (3.2.6) may be rearranged to give

$$Y = C[-\log(1-X)]^D \tag{3.3.4}$$

It is obvious that (3.3.3), (3.3.4), and (3.2.1) are all equivalent to the basic model

$$Y = CX^D \qquad\qquad (3.3.5)$$

in which X in (3.3.5) is replaced by $\log(1/X)[= -\log X]$ in (3.3.3) and by $-\log(1-X)$ in (3.3.4). Similarly, the Oswin model (3.2.7) is also equivalent to (3.3.5), with X in (3.3.5) being replaced by $X/(1-X)$ in (3.2.7). Thus, many models at first glance may appear to be different from a basic model such as (3.3.5), but they can be made equivalent by transforming the explanatory variable X in some fashion. Since X is assumed to be without error, transformations of it, such as $\log(1/X)$, as in (3.3.3), or $-\log(1-X)$, as in (3.3.4), also do not possess error.

Three of the models among (3.2.1–3.2.13) are *not*, in fact, nonlinear regression models; (3.28), (3.29), and (3.2.10) are *linear* regression models since the parameters appear linearly in those models. For example, replacing $X/(1-X)$ by X in the Iglesias and Chirife model (3.2.8) makes it equivalent to the simple straight-line model (1.1.1) of Chapter 1. Similarly, replacing $-\log(1-X)$ in (3.2.9) by X and $1/\log X$ in (3.2.10) by X also demonstrates their equivalence with the straight-line model.

The GAB model (3.2.11) fits sorption data very well over a wide range of water activities, but its statistical properties in estimation are not particularly good. However, by using the principle of reparameterization described in Section 2.3.2, we can eliminate the parameters in the numerator of (3.2.11) by dividing numerator and denominator by $Y_M Ck$ and multiplying the two terms in the denominator together to obtain

$$Y = X/(A + BX - DX^2) \qquad\qquad (3.3.6)$$

where the "new" parameters A, B, and D are functions of the "old" parameters Y_M, C, and k. If we compare (3.3.6) with the Hailwood-Horrobin model (3.2.12), we can see that they are identical, after rearrangement of (3.2.12) to allow Y to be an explicit function of X and the parameters. I have studied the statistical properties in estimation of (3.3.6) for water activity in foodstuffs and have found them to be excellent (unpublished results).

Consider now the Bradley model (3.2.3). This can be rearranged and rewritten as

$$Y = A \log(-B \log X) \qquad (3.3.7)$$

If we now take the Chen model (3.2.4) and rearrange it, we can obtain

$$Y = A \log(-B \log X + C) \qquad (3.3.8)$$

so that the Bradley model (3.2.3), despite its vastly different appearance from (3.2.4), is nothing more than a special case of the rearranged form (3.3.8) of the latter with $C = 0$. In fact, (3.2.4) is itself nothing more than a form of the well-known Gompertz model (see Sec. 5.3), except that X and Y have been interchanged. The reader thus can see the ease with which one can move from a relatively well-known model (the Gompertz) to a virtually unknown model (3.2.4) simply by reversing the roles of the explanatory and response variables. The statistical properties of (3.3.8) are reasonable good (Ratkowsky, unpublished results).

The D'Arcy-Watt model (3.2.13) is the least parsimonious equation of the thirteen models listed in Section 3.2 since it contains a total of five parameters. Despite this large number of parameters, there is no evidence that this model fits food isotherms data better than the more parsimonious models having but two or three parameters. Also, the first term of (3.2.13) is a Langmuir-type (1916) sorption expression, and I have shown (Ratkowsky, 1986b) that the estimators of the parameters of the basic Langmuir model and its extensions exhibit a high degree of nonlinear behavior, in contrast to Freundlich-type models (3.2.1 and its extensions), which are generally close-to-linear in their estimation behavior.

Sometimes models having apparently grossly different forms are still capable of fitting the data to similar degrees of accuracy. Consider Fig. 3.2, which shows a set of data for water content Y versus water activity X for casein (data courtesy of Dr. M. Kent, Torry Research Station, Aberdeen). Also drawn on Fig. 3.2, in addition to the 20 data points, are the fitted lines for the two-parameter rearranged Bradley model (3.3.7), the three-parameter rearranged Chen model (3.3.8), and the three-parameter GAB model, using the rearranged Hailwood–Horrobin form (3.3.6). From the point of view of goodness-of-fit, there is very little to distinguish between these three different models *within* the range of the data. The models differ when they are extrapolated outside the range of the data. Whereas the GAB model has $Y = 0$ when $X = 0$, as theoretically required, the Bradley and Chen equations may give infeasible results, such as negative moisture contents. Big differences between

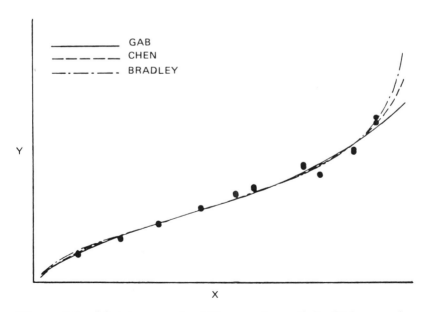

Figure 3.2   Moisture content Y vs. water activity X for casein.

the models also occur when X approaches 1. Whereas the GAB model gives a sensible result when $X = 1$ ($\hat{Y} = 21.26$), the Bradley model fails since the logarithm of zero is $-\infty$. This problem may not necessarily occur with the Chen model (3.3.8) if $C > 0$ but will cause a similar problem of indeterminacy if $C < 0$, as is the case with many food isotherms.

The users of models must therefore ask themselves whether they want merely a "good fit" to the existing data or whether there is interest in going further and trying to discover some universal relationship. It should be apparent from the wide diversity of isotherm shape for foodstuffs, as an examination of the 501 figures in Iglesias and Chirife (1982) immediately shows, that no simple model is adequate to describe each of the possible shapes. However, there may be two or three models that will serve to describe all but a few specialized cases. Iglesias and Chirife (1982) examined the fitting abilities of 9 two-parameter isotherm models, including 8 of the ones listed in Section 3.2, in an attempt to discover which of these had the best fitting properties. They fitted the 9 models to each of the data sets in their 501 figures, requiring about 9000 model/data set combinations to be fitted, as many of the figures depict more than one data set. No single equation predominated as being the best fitting equation. Presumably, they restricted themselves to two-parameter models because of the sheer volume of work that would have been involved if they had to consider three-parameter models as well. One must assume, however, that one or more of the three-parameter models would have fitted some data sets significantly better than any of the two-parameter models.

3.4 STATISTICAL VERSUS PRACTICAL CONSIDERATIONS: IS THERE A CONFLICT?

Before summarizing the results from the previous section, it is worth asking whether there are considerations other than estimation behavior that should be taken into account in deciding that a nonlinear regression model is a suitable one to use in practice. In choosing a model with good statistical properties, might that model contain parameters of no physical or biological significance? Would we not sacrifice, to some extent at least, our understanding of the physical process by using a parameterization that did not possess a readily understood interpretation? A concomitant question might be: Since parameter α in the standard model in use in our discipline is well recognized by all of us as representing (say) the point at which buildup processes are exactly balanced by breakdown processes, shouldn't we be retaining that parameter in our model even though your statistical analysis tells me that the least-squares estimator of that parameter is badly biased and nonnormally distributed?

In an attempt to answer the questions posed in the preceding paragraph, it is helpful to examine a concrete example with a specific model. We choose the so-called Michaelis–Menten model, the use of which is very widespread in biochemistry for modeling reaction rates in enzyme systems. The model may be written as

$$Y = V_{\max}X/(X + K_m) \tag{3.4.1}$$

where Y is the reaction velocity, X the substrate concentration, $V_{\max}$ a parameter representing the maximum velocity that is theoretically obtainable, and K_m a parameter that may be interpreted as the substrate concentration at which the velocity is half the maximum velocity. As data to be used in combination with (3.4.1), we choose the data set consisting of 12 pairs of X and Y values that appeared in Bates and Watts (1980), Table 1, p. 10. For these data, the measure of skewness of Hougaard (1985), expressed as g_1 (see Sec. 2.2.2), yields the following values:

Parameter	g_1
$V_{\max}$	0.989
K_m	1.180

As these values of g_1 are rather high, the estimators of the two parameters are both far-from-linear in their estimation behavior. A consequence of this is not only that the estimates of these parameters will be severely biased but also that confidence intervals for these parameters, based on the usual "linearization" method (see "Linearization Confidence Intervals" under Sec. 2.4.2), will be quite inaccurate. Table 3.1 presents 95% confidence intervals for V_{max} and K_m, calculated using expressions (2.4.7) and (2.4.8) for the likelihood and linearization methods, respectively, for the data in Bates and Watts (1980).

The agreement between the two methods is rather poor, with the likelihood limits expected to provide better "coverage" of the true but unknown parameter values than the linearization limits. The consequences of these results for the practitioner should be disquieting. That is, despite the fact that the user may deem (3.4.1) to be a suitable parameterization for studying the effect of substrate concentration on reaction velocity, the poor agreement between the likelihood and linearization intervals (and also their corresponding confidence regions, had they been calculated) must cast some doubt on the practical value of that parameterization, unless one undertakes the considerable computational effort required to calculate the likelihood (or lack-of-fit) confidence limits. Otherwise, reliance on the usual linearization limits, as produced routinely by some standard computer packages, will frequently lead to erroneous conclusions.

Table 3.1   95% Confidence Intervals for V_{max} and K_m in (3.4.1), [Data in Bates and Watts (1980)]

	V_{max}	K_m
Likelihood method	(0.076, 0.171)	(0.96, 3.60)
Linearization method	(0.066, 0.145)	(0.64, 2.76)

To illustrate how a close-to-linear parameterization can do better, consider the reparameterization of (3.4.1) given by

$$Y = X/(\theta_1 X + \theta_2) \qquad (3.4.2)$$

This parameterization was shown (Ratkowsky, 1986a) to be close-to-linear in combination with a wide range of enzyme kinetics data. Table 3.2 presents 95% likelihood and linearization confidence intervals for θ_1 and θ_2, calculated using (2.4.7) and (2.4.8), respectively, for (3.4.2) in combination with the data in Bates and Watts (1980). Although the agreement is not perfect, there is a much greater overlap of the likelihood and linearization limits than was the case with V_{max} and K_m in Table 3.1.

A class of reparameterizations that deserves to achieve wide user acceptability without sacrificing ease of interpretability is one containing only expected-value parameters (Sec. 2.3.1). An expected-value parameter is merely the fitted or predicted value corresponding to a value of the explanatory variable specified by the user. Thus, to reparameterize (3.4.1), one chooses two values of the substrate concentration, calling them X_1 and X_2, to which the fitted values y_1 and y_2 correspond. Solving for V_{max} and K_m in terms of y_1 and y_2, and substituting for those parameters in (3.4.1), yields

$$Y = X(X_2 - X_1)y_1 y_2 / [(X - X_1)X_2 y_1 + (X_2 - X)X_1 y_2] \qquad (3.4.3)$$

Table 3.2 95% Confidence Intervals for θ_1 and θ_2 in (3.4.2) [Data in Bates and Watts (1980)]

	θ_1	θ_2
Likelihood method	(5.86, 13.08)	(12.19, 21.56)
Linearization method	(5.95, 12.98)	(11.56, 20.68)

The new parameters y_1 and y_2 are interpreted as the expected (or average predicted) reaction velocities at X_1 and X_2, respectively. When the same values of X_1 and X_2 for each of several sets of data are chosen, comparisons between data sets are reduced to comparisons between expected reaction velocities at those user-chosen substrate concentrations. Table 3.3 presents 95% likelihood and linearization confidence intervals for y_1 and y_2, calculated using (2.4.7) and (2.4.8), respectively, for (3.4.3) in combination with the data of Bates and Watts (1980), for $X_1 = 0.2$ and $X_2 = 2.0$, these substrate concentrations being the lowest and the highest, respectively, in the data set. Clearly, there is excellent agreement between the confidence intervals calculated in the two different ways. These results confirm the assertion made in Section 2.4 that the more close-to-linear the model, the greater the agreement between the model's linearization and likelihood confidence sets. In contrast, a far-from-linear model has been shown to have poor agreement between the linearization and likelihood confidence sets.

The information in the preceding paragraphs provides a justification for seeking parameterizations that are close-to-linear. With a close-to-linear parameterization, the user can feel assured that the confidence intervals reported by standard packages, obtained using the linearization method, will be "reasonable" ones. Many users of nonlinear regression models will be quite satisfied with an empirically determined model. The standard of judgment that will often be applied to these models will be based on considerations such as (1) whether the model fits the observed data

Table 3.3   95% Confidence Intervals for y_1 and y_2 in (3.4.3), [Data in Bates and Watts (1980), with $X_1 = 0.2$ and $X_2 = 2.0$]

	$100\,y_1$	$100\,y_2$
Likelihood method	$(0.875, 1.370)$	$(5.012, 6.404)$
Linearization method	$(0.863, 1.359)$	$(5.008, 6.405)$

well, (2) whether the parameter estimates in the reparameterized model are reasonably well determined (that is, whether their
standard errors are reasonably small compared with the estimates
themselves), and (3) whether comparisons between different sets
of data, usually based on the parameter estimates, can be readily made. Empirically determined close-to-linear models, such as
those containing expected-value parameters, will satisfy all these
desiderata, thereby fulfilling the requirements of most users of
nonlinear regression models.

Let us consider next those modeling situations for which there
is a "standard" model, based on either theoretical considerations or long-established practice, for which an empirical model,
such as one incorporating expected-value parameters, will not be
deemed by modelers to be a satisfactory alternative. Under these
circumstances, the modeler will wish to retain the standard parameterization and to report parameter estimates and confidence
sets for the parameters of that parameterization. As we have seen
from the results presented earlier in this section, the standard parameterizations often have parameters whose estimators exhibit
far-from-linear behavior, so that confidence sets based on the linearization method may be grossly misleading. The modeler will
then have no alternative, if accurate confidence statements are to
be made, to calculating confidence sets based on either the likelihood or lack-of-fit approaches (Sec. 2.4). This may involve the
modeler in considerable programming effort since, as has already
been stated, only the linearization method is widely available in
standard statistical packages. Far-from-linear models will tend to
have very asymmetric confidence sets; their parameters are usually not well determined (the ratios of the parameter estimates to
their standard errors often being rather low). In addition, comparisons between different sets of data pose considerable practical
difficulties (see, for example, Ratkowsky, 1983, Sec. 7.4). Indeed,
one of the arguments advanced by the proponents of the concept
of retaining a standard parameterization is that workers in that
discipline are familiar with the parameterization in question and
have some "feeling" for the magnitude of the parameter values
that may be expected in typical situations. This feeling is supposed

to assist in the making of comparisons between data sets derived under differing experimental conditions. However, it must not be forgotten that if a nonlinear regression model is far-from-linear, as is very often the case for many standard parameterizations in common use, then (1) the parameter estimates may be very biased, and (2) the standard errors of the parameter estimates may be grossly underestimated. This makes comparisons between estimates derived from (say) two sets of experimental conditions very dubious since the extent of the biases in the two sets of experiments is, in general, different. Unlike the case of comparing unbiased (or approximately unbiased) estimators, each having a normal (or approximately normal) distribution, with possibly different means and different variances, comparisons of biased estimators, each having an unknown nonnormal distribution, are impractical. Thus, one of the advantages of a standard parameterization, that of user familiarity with typical values of parameter estimates, may be lost. Of course, one can counteract far-from-linear behavior by opting for large experiments since the greater the sample size, the closer-to-linear the estimation behavior. However, the larger the experiment, the more costly the experiment, and cost may be a prohibitive factor. In any event, one of the virtues of a close-to-linear parameterization, such as one obtains with expected-value parameters, is that its asymptotic properties, namely, unbiasedness, minimum variance, normally distributed estimators, is closely approximated in a close-to-linear model, *even for the relatively small sample sizes that are dictated by affordable experimentation.* Thus, it is this author's opinion that there are cogent reasons for preferring close-to-linear models, even in disciplines where certain "standard" parameterizations have become established. In many of these cases, there was no real scientific basis for selecting that particular parameterization anyway. Often, the choice of parameterization was an arbitrary one that appeared in an influential paper by one of the workers in a particular discipline, and subsequent authors simply followed suit uncritically.

It is, in fact, very difficult to provide an appropriate scientific justification for virtually any parameterization. Many models can be derived as solutions to differential equations, but such solutions will have at least one arbitrary constant, which results in the parameter appearing in an arbitrary form. This parameter can always be reparameterized without altering the underlying scientific basis of the model. Furthermore, the same scientific principles may be expressed by alternative forms of differential equations, leading to different parameterizations. To illustrate these points, consider the following differential equation:

$$dY/dX = \beta Y \qquad (3.4.4)$$

Solving, one obtains

$$\log Y = \beta X + c$$

where c is an arbitrary constant. If one writes this constant as $\log \alpha$, then a familiar nonlinear regression model,

$$Y = \alpha \exp(\beta X) \qquad (3.4.5)$$

results. However, $\log \alpha$ was an arbitrary choice for c, as is any other choice. Another choice could have been to let $c = \gamma$, in which case, one would have obtained

$$Y = \exp(\beta X + \gamma) \qquad (3.4.6)$$

The choice is, of course, unlimited. Similarly, the constant β in the differential equation (3.4.4) serves only the purpose of indicating that the rate at which Y changes with respect to X (that is, dY/dX) is proportional to Y. This idea could have been expressed just as readily by $\log \lambda$, where λ is another constant, rather than by β itself. The equivalent differential equation

$$dY/dX = (\log \lambda)Y \qquad (3.4.7)$$

has as its solution

$$\log Y = (\log \lambda)X + c = \log \lambda^X + c$$

Letting $c = \log \alpha$, although, as before, an unlimited choice is possible, one obtains

$$Y = \alpha \lambda^X \qquad\qquad (3.4.8)$$

It is obvious that (3.4.8) is merely a reparameterization of (3.4.5), illustrating that arbitrariness in the way a differential equation is written down is mirrored by a corresponding arbitrariness in the form of the ensuing parameterization. This lack of uniqueness provides a justification for the use of alternative parameterizations. Since the user is fixed not by the choice of parameterization but only by the choice of the basic underlying model, there is no justification for not seeking parameterizations with close-to-linear estimation behavior.

3.5 SUMMARY OF APPROPRIATE STEPS TO FIND A SUITABLE REGRESSION MODEL FOR ONE'S DATA

The following remarks are made in an attempt to summarize some of the goals and principles a modeler should have in mind when trying to find an appropriate model to relate the response variable Y to a set of explanatory variables $X_1, X_2, ..., X_p$, where p can be, and usually is, as little as 1 when there is only a single explanatory variable. The parameters, usually indicated by Greek letters or uppercase Roman letters, are the coefficients to be estimated.

(1) Decide what the correct error distribution is for the dependent variable. Search for the model having the correct stochastic assumption with the response variable expressed explicitly as a

function of the explanatory variables. Models (3.2.1), (3.2.7–3.2.11), and (3.2.13) all have the response variable expressed as a function of X. The other models in Section 3.2 do not. Whether an "additive" error assumption is appropriate depends on the data sets and knowledge of the physical (biological, etc.) systems being modeled. A multiplicative error assumption may be introduced by taking the logarithm of both sides of the model as in (3.3.2). The development of generalized linear models (Nelder and Wedderburn, 1972), as realized by the statistical package GLIM (1978), makes it possible to consider other stochastic assumptions, such as gamma or Poisson error. In addition, some nonlinear regression models having the usual normal error assumption may also be fitted using this package.

(2) Before committing oneself to a nonlinear regression model, a linear regression model should be tried first. Linear models have the advantage that the least-squares estimators of their parameters (with independent and identically distributed errors) are unbiased, minimum variance, normally distributed estimators. Of course, this is true only if the chosen model is the true model. If the model fitted is not the true model, then the least-squares estimators of the parameters do not have these desirable properties in estimation. Indeed, it becomes difficult even to speak about the properties of the estimators of a parameter that is present in an incorrect model. In general, one should avoid artificial linear models such as the simple polynomials, namely,

$$Y = \alpha + \beta X + \gamma X^2 + \dots \tag{3.5.1}$$

since such models seldom arise naturally in scientific work and are rarely solutions of differential equations. With several explanatory variables X_1, X_2, ..., X_p, the user may be tempted to use a multiple variable linear regression model (see, for example, Gunst and Mason, 1980). This often results in satisfactory models for prediction of the response variate but may at times give grossly misleading results (Miller, 1984). It should always be remembered that such models are coarse approximations to the true models, which are likely to be nonlinear.

(3) Nonlinear regression models should be parsimonious. One should not use a model with three parameters where a model with two parameters would suffice (see Sec. 2.5.4). In general, the degree of nonlinear behavior tends to increase with the number of parameters, and there are very few models with more than four parameters that exhibit close-to-linear behavior. The object of nonlinear regression modeling should be to find a parsimonious model exhibiting close-to-linear behavior. It may not be the "true" model, as such a model may never be found, but if its parameters are interpretable in some way, the model can find use in practice.

(4) Recognize that many equations are really the same basic model but appear in different forms or guises called parameterizations. Choose a parameterization that has good estimation behavior (that is, one that is close-to-linear). Expected-value parameters are likely to have better estimation behavior than any other type of parameter. Thus, whenever possible, one should replace the parameters giving rise to poor behavior of their least-squares estimators with expected-value parameters.

(5) Models such as (3.2.7) and (3.3.3–3.3.5) were seen to be the same basic model, in which the explanatory variable was transformed (in each case by a different transformation). Transformations may be successful in other cases as well. For example, in dose-response studies where increasing concentrations of an insecticide results in an increasing kill rate, it is usually found that the cumulative response, proportion killed, is well fitted by a logistic

$$\text{proportion killed} = 1/[1 + \exp(\beta - \gamma X')] \qquad (3.5.2)$$

or by a probit (cumulative normal distribution) where the logarithm of the dose X is used (that is, $X' = \log X$). Similarly, in bioassay, the response may be a logistic curve

$$Y = \alpha/[1 + \exp(\beta - \gamma X')] + \delta \qquad\qquad (3.5.3)$$

where X' is the logarithm of the concentration of the assayed chemical agent or compound rather than the concentration itself.

II

The Models

INTRODUCTION

The first part of this handbook was devoted to the background material and considerations that a modeler must undertake to develop an appropriate model for a system under study. This second part is devoted to the presentation and discussion of the models themselves. The arrangement of the presentation of the models is such that each chapter is devoted to models having a particular shape, given that they have a certain number of explanatory variables. Thus, Chapters 4–6 all deal with models having a single explanatory variable X, whereas Chapter 7 deals with models with two or more explanatory variables. Chapter 4 concerns itself with convex/concave curves, that is, ones that are continuously ascending or descending without maxima, minima, or inflection points. Chapter 5 deals with curves possessing inflection points but without maxima or minima. Such curves may have asymptotes at $X = -\infty$ or $X = +\infty$, or they may have maxima or minima at infeasible values of X, such as at a negative value of X for problems in which only positive X values are sensible. The models of Chapter 5 are thus typically sigmoidal-shaped curves. Chapter 6 concerns itself with models having one or more maxima or minima and may, and usually do, have one or more inflection points. Hence, it is clear that these chapters have been organized with respect to the shape or form of the curve rather than the use to which the models may be put. This decision flows from the belief that the same-shaped model may be used for widely differing applications in different disciplines.

It is important to realize that the statistical properties of a model remain the same, irrespective of the discipline or application of the model. Also, it should be clear that any nonlinear regression model will exhibit close-to-linear estimation behavior if the sample size (that is, the number of data points) is sufficiently large. Sometimes, the sample size required to achieve close-to-linear behavior is far greater than would be practically possible. Hence, the expression "close-to-linear" is used in this handbook to refer to nonlinear regression models whose least-squares estimators come close to approximating the large-sample properties of such estimators in the relatively small sample sizes used in practice, sometimes with as few as a half-dozen data points. Such well-behaving models are marked with an asterisk (*) in this handbook. A model marked with an asterisk is not necessarily the best possible parameterization of that model. One would usually be content to find a parameterization for which the bias of the estimators is small enough for all practical purposes. Concomitantly, the estimators will be close to having a normal distribution and to having the minimum possible variance.

Sometimes, a parameterization may be found, usually as a result of an expected-value transformation (see Sec. 2.3.1), which is even closer-to-linear than some other otherwise acceptable close-to-linear parameterization having a simple equation form. In this handbook, these "super" close-to-linear models are indicated by a double asterisk (**). Because models containing expected-value parameters are usually more complicated algebraically, sometimes considerably more so, than other types of models, the reader may prefer to choose the simpler model marked with a single asterisk if such a model is available. In a few cases in which the model under consideration is in fact a linear model, a triple asterisk (∗∗∗) is used.

Generally, with only a few exceptions, the models presented in Chapters 4–7 will be written with a lowercase y on the left-hand side of the equations. This shorthand for the "expectation of Y" (Sec. 1.4.1) implies an *additive error* assumption, meaning that the stochastic term should be added to the expression on the right-hand side of the equation. In a few cases in which a *multiplicative error* assumption is discussed, the logarithm of the right-hand side of the expression must be taken, with the left-hand side written $E(\log Y)$. The stochastic term should be added to the right-hand side of the equation *after* the logarithm has been taken. This is the same as if the error were multiplied by the expression on the right-hand side *before* logarithms are taken.

Chapter 8 considers other models that do not fall into the categories delimiting each of Chapters 4–7. This chapter also deals with some of the models that have been excluded from consideration in this handbook, with reasons presented for their exclusion.

Chapter 9 deals with the question of how to obtain initial parameter estimates as starting values for use with algorithms or procedures for obtaining the least-squares estimates. The final chapter (Chapter 10) offers a summary or overview of the whole handbook, describing considerations about models in different chapters or in different sections of a chapter. Some "rules" or principles, which could be overlooked if one thought only about the individual models in isolation, are advanced.

4

Models with One X Variable, Convex/Concave Curves

This chapter concerns models with a single explanatory variable in which the graph of Y versus X is either an ascending or descending convex or concave curve. In mathematical parlance, a function $f(X)$ is said to be convex if the second derivative of the function with respect to X is greater than zero for all X. A concave function is one in which the second derivative is less than zero for all X. Typically, then, the models in this chapter will not have maxima, minima, or inflection points.

4.1 ONE-PARAMETER CURVES

Among the wide variety of simple curves containing but a single parameter (see Fig. 4.1) is

75

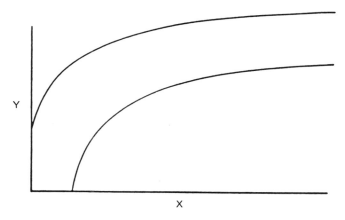

Figure 4.1

(∗) $y = \log(X - \alpha)$ (4.1.1)

The asterisk at the left-hand margin alongside (4.1.1) indicates
that the statistical properties in estimation of that model are good,
that is, that the model behaves in a reasonably close-to-linear
manner in estimation. An even better-behaved model is obtained
by replacing α by an expected-value parameter, to yield

(∗∗) $y = \log[X - X_1 + \exp(y_1)]$ (4.1.2)

where y_1 is the expected (or fitted) value corresponding to $X = X_1$,
where X_1 should be chosen to be somewhere within the observed
range of the X values in the data set. For example, $\bar{X}$, the sample
mean of the Xs, would be a suitable choice for X_1, but the exact
value of X_1 is not critical provided that it does not fall outside the
observed range.
     Another simple curve with one parameter (see Fig. 4.2a,b) is

(∗) $y = 1/(1 + \alpha X)$ (4.1.3)

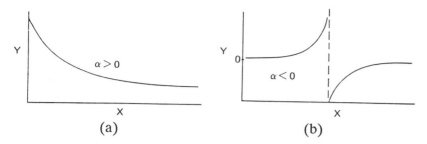

Figure 4.2

When $\alpha < 0$, there is a vertical asymptote occurring at $X = -1/\alpha$. (Since α is negative, this corresponds to a positive value of X.) Replacing α with an expected-value parameter y_1, corresponding to $X = X_1$, results in

$$(**) \quad y = 1/[1 + (X/X_1)(1/y_1 - 1)] \tag{4.1.4}$$

Readers can readily verify, for this and for other models with expected-value parameters, that when $X = X_1, y = y_1$.

Another single-parameter simple curve (see Fig. 4.3) is

$$(*) \quad y = \exp(X - \alpha) \tag{4.1.5}$$

This model is, in fact, a disguised intrinsically linear model, since (4.1.5) may be reparameterized to yield a linear model. That is, replacing α by an expected-value parameter y_1, corresponding to $X = X_1$, yields

$$(***) \quad y = y_1 \exp(X - X_1) \tag{4.1.6}$$

which is clearly linear in the parameter y_1.

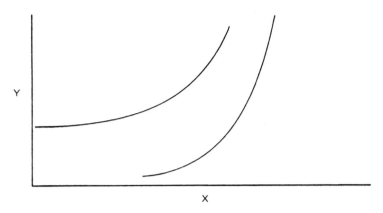

Figure 4.3

Yet another simple curve with one parameter (see Fig. 4.4) is

$$(*) \quad y = X^{\alpha} \tag{4.1.7}$$

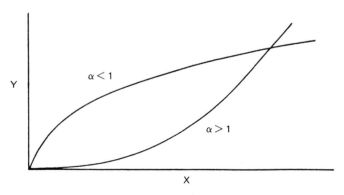

Figure 4.4

The parameter α may be replaced by an expected-value parameter y_1, corresponding to $X = X_1$, to yield

$$(**) \quad y = y_1^{\log X / \log X_1} \tag{4.1.8}$$

A further simple one-parameter curve (see Fig. 4.5a,b) is

$$y = 1/(X + \alpha) \tag{4.1.9}$$

The parameter α may be replaced by an expected-value parameter y_1, corresponding to $X = X_1$, to yield

$$(**) \quad y = 1/(X - X_1 + 1/y_1) \tag{4.1.10}$$

Yet another single-parameter simple curve (see Figure 4.6a–c) is

$$(*) \quad y = 1 - X^{-\alpha} = 1 - 1/X^{\alpha} \tag{4.1.11}$$

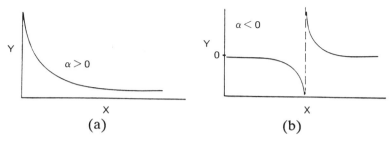

(a)   (b)

Figure 4.5

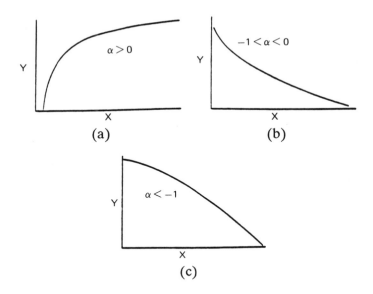

Figure 4.6

which is based on the Pareto distribution. Replacing α by an expected-value parameter y_1, corresponding to $X = X_1$, results in

$$(**) \quad y = 1 - (1 - y_1)^{\log X / \log X_1} \tag{4.1.12}$$

A further simple one-parameter curve (see Fig. 4.7) is

$$(*) \quad y = \alpha \exp(-\alpha X) \tag{4.1.13}$$

the single parameter α appearing in two places. A reparameterization of the above, obtained by replacing α by $\beta = \exp(-\alpha)$, is

$$(*) \quad y = -(\log \beta) \beta^X \tag{4.1.14}$$

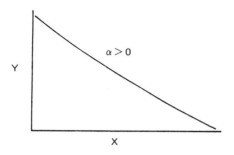

Figure 4.7

This has very similar estimation properties to (4.1.13).
   Another one-parameter simple curve (see Fig. 4.8a,b) is

$$(*) \quad y = \exp(-\alpha X) \tag{4.1.15}$$

which is similar to (4.1.13) but without the parameter appearing
twice. A slightly better parameterization is

$$(*) \quad y = \beta^X \tag{4.1.16}$$

but the closest-to-linear estimation behavior results from the

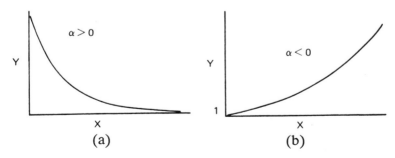

Figure 4.8

by an expected-value parameter y_1, corresponding to $X = X_1$, to give

$(\ast\ast) \quad y = y_1^{X/X_1}$ \hfill (4.1.17)

The simple one-parameter model

$(\ast) \quad y = 1 - \exp(-\alpha X)$ \hfill (4.1.18)

is identical, in estimation behavior, to that of (4.1.15). Similarly,

$(\ast) \quad y = 1 - \beta^X$ \hfill (4.1.19)

has identical estimation properties to (4.1.16). The best behavior is obtained from the expected-value parameterization

$(\ast\ast) \quad y = 1 - (1 - y_1)^{X/X_1}$ \hfill (4.1.20)

The graphs for models (4.1.18–4.1.20) are the same as those for (4.1.15–4.1.17) (see Figure 4.8a,b) except that they are reflected through the horizontal line $y = 1$.

The models covered in this chapter clearly do not exhaust all the possibilities of one-parameter convex or concave curves. Some models which are capable of exhibiting a sigmoidal shape may be convex or concave for restricted values of the parameter. For example, (5.1.1) of Chapter 5 with $\alpha < 1$ results in a curve which is concave towards the x-axis.

4.2 TWO-PARAMETER CURVES

A model such as (4.1.1) can be made into a two-parameter model
by multiplying its right-hand side by another parameter to give

$$(*) \quad y = \beta \log(X - \alpha) = \log(X - \alpha)^{\beta} \tag{4.2.1}$$

The shape of (4.1.1) has not been changed by this multiplication
(see Figure 4.9a,b), but (4.2.1) is more general than (4.1.1) as the
response variable y is now scaled. Either α or β in (4.2.1) may be
replaced by an expected-value parameter. Although this is often
not necessary, as the estimation behavior of (4.2.1) is quite close-
to-linear, the worse-behaving parameter of the two is usually α and
may be replaced by y_1, corresponding to $X = X_1$, to give

$$(**) \quad y = \beta \log(X - X_1 + \exp(y_1/\beta)) \tag{4.2.2}$$

Another two-parameter curve (see Fig. 4.10a–c) is

$$y = \log(\alpha + \beta X) \tag{4.2.3}$$

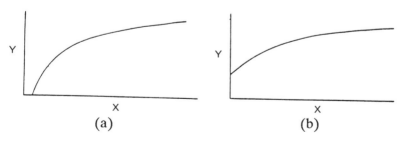

Figure 4.9

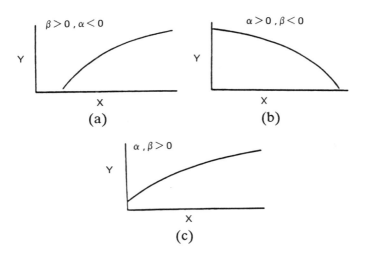

Figure 4.10

The estimation properties of (4.2.3) are considerably improved if both parameters α and β are replaced by expected-value parameters to give

$$(**)\quad y = \log\{[(X_2 - X)/(X_2 - X_1)]\exp(y_1)$$
$$+ [(X - X_1)/(X_2 - X_1)]\exp(y_2)\}\quad (4.2.4)$$

Here, y_1 is the expected-value parameter corresponding to $X = X_1$, and y_2 is the expected-value parameter corresponding to $X = X_2$. X_1 and X_2 may be chosen to be the lowest and highest observed X values, respectively, in the data set, or they may be any other values within the range of the data set, provided that they are not too close together, in which case the parameters y_1 and y_2 become too highly correlated with each other.

A further model with two parameters (see Fig. 4.11) is

$$y = \alpha(1 - \beta^X)\quad (4.2.5)$$

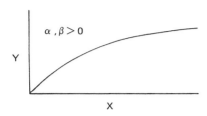

Figure 4.11

Model (4.2.5) is better than

$$y = \alpha[1 - \exp(-\gamma X)] \tag{4.2.6}$$

where $\gamma = -\log(\beta)$. However, α in (4.2.5) can be replaced by an expected-value parameter to yield

$$(*) \quad y = (1 - \beta^X)y_1/(1 - \beta^{X_1}) \tag{4.2.7}$$

where y_1 corresponds to $X = X_1$. The parameter β cannot be replaced by an expected-value parameter, y_2 (say), corresponding to $X = X_2$, to yield an expression for y in terms of y_1 and y_2, because β cannot be algebraically expressed in terms of y_1 and y_2, *in general*. However, if one chooses X_1 and X_2 such that X_2 is exactly twice X_1, the term $1 - \beta^{2X_1}$ can be factored into

$$(1 - \beta^{X_1})(1 + \beta^{X_1})$$

allowing the denominator of the resulting expression to cancel the first term in the numerator, yielding, after substitution of this solution back into (4.2.7),

$$(**) \quad y = \{1 - [(y_2/y_1) - 1]^{X/X_1}\}/[2 - (y_2/y_1)] \tag{4.2.8}$$

It must be remembered that (4.2.8) is valid only when X_1 and X_2 are chosen so that $X_2 = 2X_1$. If it is not possible to choose X_1 and X_2 so that both of them fall within the range of the observed X values in the data set, the estimation properties of (4.2.8) will not be as good as when they do fall within the range.

Another two-parameter model (see Fig. 4.12a,b) is

$$y = \alpha \beta^X \qquad\qquad (4.2.9)$$

Alternative parameterizations of (4.2.9) are the following

$$y = \alpha \exp(\gamma X) \qquad\qquad (4.2.10)$$
$$(*) \quad y = \exp(\delta + \gamma X) \qquad\qquad (4.2.11)$$

where $\gamma = \log \beta$ and $\delta = \log \alpha$. The statistical properties of (4.2.11) are better than those of (4.2.9) and (4.2.10). In addition, (4.2.11) is a *generalized linear model* and therefore may be fitted using the computer package GLIM, for example. The *linear predictor* is $\delta + \gamma X$, which is related to y via a *log link*. The parameters in (4.2.9–4.2.11) may be replaced by expected-value parameters to yield

$$(**) \quad y = y_1^{(X-X_2)/(X_1-X_2)} y_2^{(X_1-X)/(X_1-X_2)} \qquad\qquad (4.2.12)$$

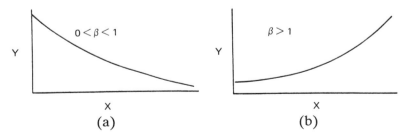

(a)   (b)

Figure 4.12

A further two-parameter model (see Fig. 4.13a–c) is

$$(*) \quad y = \alpha X^{\beta} \tag{4.2.13}$$

which is the classical Freundlich (1926) model. Sometimes, the parameterization

$$y = \alpha X^{1/\gamma} \tag{4.2.14}$$

is used in preference to (4.2.13), but its statistical properties are no better than those of that model. The parameters α and β in (4.2.13) may be replaced by expected-value parameters y_1 and y_2, corresponding to $X = X_1$ and $X = X_2$, respectively, to yield

$$(**) \quad y = y_1(y_2/y_1)^{\log(X/X_1)/\log(X_2/X_1)} \tag{4.2.15}$$

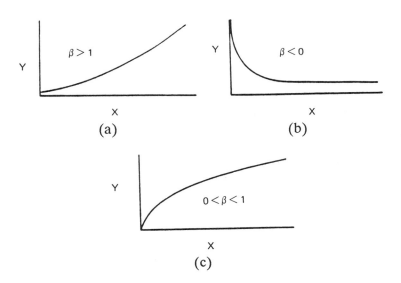

Figure 4.13

a parameterization with excellent statistical properties in estimation.

Another commonly used two-parameter model (see Figure
4.14) is the rectangular hyperbola

$$y = \alpha\beta X/(1 + \beta X) \tag{4.2.16}$$

known often as the simple Langmuir model. A reparameterization
of (4.2.16),

$$y = y_{\max}X/(X + X_K) \tag{4.2.17}$$

where $y_{\max}$ and X_K are its parameters, known in enzyme kinetics
as the Michaelis–Menten model, has, in common with (4.2.16),
rather poor estimation properties (Ratkowsky, 1986a). Instead, a
good substitute for (4.2.16) and (4.2.17) is

$$(*) \quad y = X/(\theta_1 X + \theta_2) \tag{4.2.18}$$

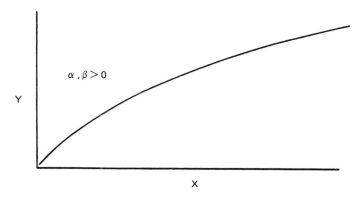

Figure 4.14

which can also be rewritten as

$$y = 1/(\theta_1 + \theta_2/X)$$

demonstrating more clearly that it is a generalized linear model with linear predictor $\theta_1 + \theta_2/X$ related to y via a reciprocal link. Hence, it may be fitted using GLIM. An expected-value parameterization of (4.2.16–4.2.18) is also available, having the form

$$(**) \quad y = X(X_2 - X_1)y_1y_2/[(X - X_1)X_2y_1 + (X_2 - X)X_1y_2]$$

(4.2.19)

which has excellent estimation properties.
    A further two-parameter model (see Fig. 4.15a–c) is

$$(*) \quad y = 1/(\alpha + \beta X) \qquad\qquad (4.2.20)$$

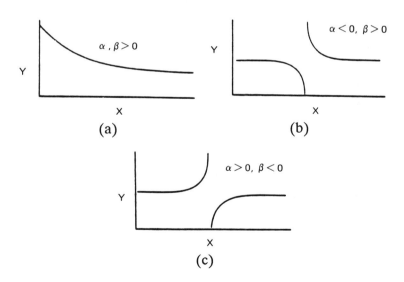

(a)   (b)

(c)

Figure 4.15

In yield-density studies in agricultural research, this is known as the reciprocal model (Shinozaki and Kira, 1956). Another parameterization of (4.2.20) is

$$y = \gamma/(1 + \delta X) \tag{4.2.21}$$

where $\gamma = 1/\alpha$ and $\delta = \beta/\alpha$, but this is slightly inferior to (4.2.20). The form of (4.2.20) indicates that it is a generalized linear model with linear predictor $\alpha + \beta X$ and a reciprocal link. It can therefore be fitted using GLIM. The parameters in (4.2.20) and (4.2.21) can be replaced by expected-value parameters y_1 and y_2, corresponding to $X = X_1$ and $X = X_2$, respectively, to yield

$$(**) \quad y = y_1 y_2 (X_2 - X_1)[y_2(X_2 - X) + y_1(X - X_1)] \tag{4.2.22}$$

which has excellent estimation properties.
   Another two-parameter model (see Fig. 4.16) is

$$(*) \quad y = \alpha(X - \beta) \tag{4.2.23}$$

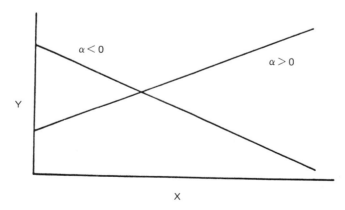

Figure 4.16

This is really a reparameterization of the straight-line model
(1.1.2), which has been used for fitting special cases of the
Bělehrádek (1935) model. For example, in the "square root"
model (Ratkowsky et al., 1982), X is absolute temperature, y the
square root of the growth rate constant, and β a notional minimum temperature for growth or "biological zero." Of course, one
can use the straight-line model (1.1.2) directly since that model exhibits no nonlinear behavior at all. However, the form of (4.2.23)
is very convenient when one wishes to interpret β as the minimum
notional temperature.

   Yet another two-parameter model (see Fig. 4.17a–c) is

$$(*) \quad y = 1 - 1/(1 + \alpha X)^{\beta} \qquad\qquad (4.2.24)$$

The statistical properties of (4.2.24) are reasonably good. The
worst-behaving parameter is β, which can be replaced by an

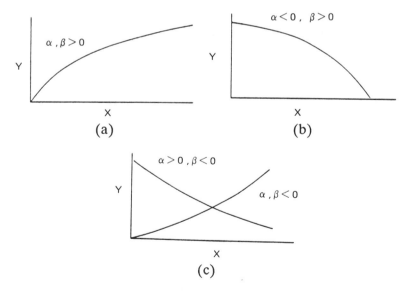

(a)   (b)

(c)

Figure 4.17

expected-value parameter y_1, corresponding to $X = X_1$, to yield

$$(**) \quad y = 1 - 1/[1/(1-y_1)]^{\log(1+\alpha X)/\log(1+\alpha X_1)} \qquad\qquad (4.2.25)$$

which, in spite of its cumbersome appearance, has better estimation properties than (4.2.24).

In Chapter 3, the rearranged Bradley (1936) model (3.3.7) (see Fig. 4.18)

$$(*) \quad y = \alpha \log(-\beta \log X)$$

was considered for use as a model relating moisture content to water activity in foodstuffs. The model can be rewritten as

$$y = \alpha \log(-\beta) + \alpha \log(\log X)$$

It is obvious that this can be reparameterized to a model that is linear in the transformed explanatory variable $\log(\log X)$ by replacing $\alpha \log(-\beta)$ by a new parameter (say) γ.

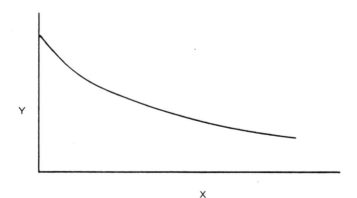

Figure 4.18

Another model that appeared in Chapter 3, also used to relate moisture content to water activity, was the BET model (3.2.2) (see Figure 4.19). That expression can be rearranged and written as

$$y = \alpha\beta X/[1 + (\beta - 2)X - (\beta - 1)X^2] \qquad (4.2.26)$$

Investigation of the statistical properties of (4.2.26) shows that α has good estimation properties but β does not. Reparameterization of (4.2.26) by dividing numerator and denominator by $\alpha\beta$, following the parameters in denominator principle (Sec. 2.3.2), yields

$$(*) \quad y = X/[\gamma + \delta X - (\gamma + \delta)X^2] \qquad (4.2.27)$$

The statistical properties in estimation of γ and δ are very similar, both being good.

Another two-parameter curve (see Figure 4.20a–c) is

$$y = \alpha(1 - 1/X^\beta) = \alpha(1 - X^{-\beta}) \qquad (4.2.28)$$

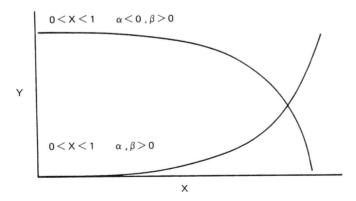

Figure 4.19

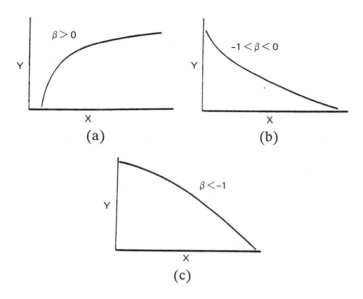

Figure 4.20

The parameter most in need of reparameterization is α. Replacing it by an expected-value parameter y_1, evaluated at $X = X_1$, yields

$$(*) \quad y = y_1(1 - X^{-\beta})/(1 - X_1^{-\beta}) \tag{4.2.29}$$

Yet another two-parameter curve (see Figure 4.21a–c) is

$$y = \alpha(1 + X)^\beta \tag{4.2.30}$$

Since α tends to have worse estimation properties than β, that parameter can be replaced by an expected-value parameter y_1, evaluated at $X = X_1$, to yield

$$(*) \quad y = y_1[(1 + X)/(1 + X_1)]^\beta \tag{4.2.31}$$

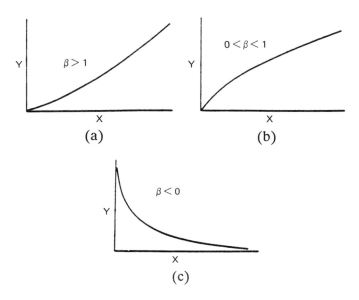

Figure 4.21

which has reasonably good estimation properties.

Other two-parameter models exist which are capable of exhibiting a sigmoidal shape but which may be convex or concave for restricted values of one or both of the parameters. Thus, for example, (5.2.1) of Chapter 5 with $\beta < 1$ results in a curve which is concave towards the x-axis.

4.3 THREE-PARAMETER CURVES

One of the most widely used curves in applied scientific work is the *asymptotic regression* model, also known in agricultural research as the Mitscherlich law and in fisheries research as the von Bertalanffy law. The basic shape is given by Figure 4.22. Various parameterizations have been used or suggested, among which are

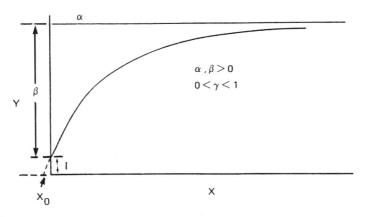

Figure 4.22

the following:

$$y = \alpha - \beta \gamma^X \qquad (4.3.1)$$

$$y = \alpha - \beta \exp(-KX) \qquad (4.3.2)$$

$$y = \alpha\{1 - \exp[-K(X - X_0)]\} \qquad (4.3.3)$$

$$y = \alpha + (I - \alpha)\gamma^X \qquad (4.3.4)$$

$$y = \alpha + (I - \alpha)\exp(-KX) \qquad (4.3.5)$$

$$y = \alpha - \exp[-(\delta + KX)] \qquad (4.3.6)$$

$$y = I + \beta[1 - \exp(-KX)] \qquad (4.3.7)$$

In (4.3.1–4.3.7), α is the asymptote corresponding to $X \to \infty$, $\beta (= \alpha - I)$ the range of the response between $X = 0$ and $X = \infty$, I the intercept of the y-axis (that is, the value of Y when $X = 0$), δ the natural logarithm of the range, and X_0 the "true" origin of the explanatory variable X (that is, the value of X when $Y = 0$); K and γ relate to the rate at which Y changes from its initial value I (at $X = 0$) to its final value α. As shown by Ratkowsky (1986c), none of the parameterizations (4.3.1–4.3.7) consistently has sufficiently good estimation properties. A parameterization with excellent statistical properties is a generalization of one put forward by Schnute and Fournier (1980),

$$(*)\quad y = y_1 + (y_2 - y_1)(1 - k^{m-1})/(1 - k^{n-1}) \tag{4.3.8}$$

where $m - 1 = (n - 1)(X - X_1)/(X_2 - X_1)$, n is the sample size (number of data pairs), and the expected-value parameters y_1 and y_2 correspond to $X = X_1$ and $X = X_2$, respectively. These X values may be taken to be the minimum and maximum observed values of X in the data set, or they may be any other values of X within the observed range of the data, provided they are not too close together, in which case y_1 and y_2 become too highly correlated with each other. The third parameter k, in (4.3.8), is related to K and γ in (4.3.1–4.3.7) by the following expressions:

$$K = -(n-1)(\log k)/(X_2 - X_1)$$

and

$$\gamma = k^{(n-1)/(X_2 - X_1)}$$

The reason why parameter k appears in (4.3.8) in preference to similar expressions involving K or γ is that the statistical properties in estimation of k are better than those of K and γ (Ratkowsky, 1986c, Table 2). However, as pointed out by Ross (1978) and by Francis (1988b), k may be replaced by a third expected-value parameter y_3, corresponding to $X = X_3$, provided X_3 is chosen to be the arithmetic mean of X_1 and X_2. Under this mild restriction, k in (4.3.8) may be eliminated to produce the following model containing only expected-value parameters:

$$(**)\quad y = y_1 + \frac{(y_2 - y_1)\{1 - [(y_2 - y_3)/(y_3 - y_1)]^q\}}{1 - [(y_2 - y_3)/(y_3 - y_1)]^2} \tag{4.3.9}$$

where $q = 2(X - X_1)/(X_2 - X_1)$ and y_1, y_2, and y_3 correspond to $X = X_1$, $X = X_2$, and $X = (X_1 + X_2)/2$, respectively. The statistical properties of (4.3.9) are excellent, and this model may be used generally whenever the asymptotic regression model is an appropriate curve for the data in a given problem. In common with other expected-value parameters, initial estimates of y_1, y_2, and y_3 are readily obtainable by reading approximate fitted values from a smooth curve through the data.

Another three-parameter curve (see Figure 4.23a–c) is an extended Langmuir model (Sibbesen, 1981)

$$y = \alpha\beta X^{1-\gamma}/(1 + \beta X^{1-\gamma}) \tag{4.3.10}$$

The estimator of γ in (4.3.10) is often close-to-linear, but those of α and β may not be. If there are no data points in the vicinity of the asymptote α, the behavior of the estimator of α can be very far-from-linear. Model (4.3.10) may be reparameterized by dividing

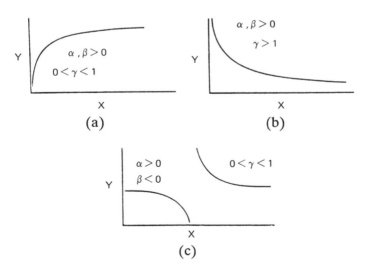

Figure 4.23

numerator and denominator by $\alpha\beta$ (and also by $X^{1-\gamma}$) to give

$$y = 1/(\theta_1 + \theta_2 X^{\gamma-1}) \tag{4.3.11}$$

The behavior of θ_1 and θ_2 in estimation is somewhat better than that of α and β in (4.3.10), but this basic model should be used with caution. Either α or β may be replaced by an expected-value parameter, but there is no explicit expression when one tries to replace both simultaneously.

An additional curve with three parameters (see Figure 4.24a–c) is the modified Freundlich model

$$y = \alpha + \beta X^{\gamma} \tag{4.3.12}$$

Although (4.3.12) is a reasonably close-to-linear model, some of its parameters, but not all, may be replaced by expected-value

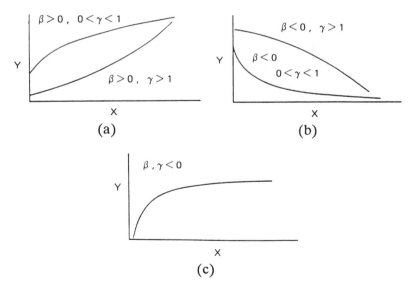

Figure 4.24

parameters. When α is the worst-behaving parameter, it can be eliminated to yield

$$(*) \quad y = y_1 + \beta(X^\gamma - X_1^\gamma) \tag{4.3.13}$$

When β and γ are worse-behaving than α, they may be eliminated to yield

$$(*) \quad y = \alpha + (y_1 - \alpha) \left(\frac{y_2 - \alpha}{y_1 - \alpha}\right)^{\log(X/X_1)/\log(X_2/X_1)} \tag{4.3.14}$$

Another three-parameter model (see Figure 4.25) is the extended Freundlich model (Sibbesen, 1981)

$$(*) \quad y = \alpha X^{\beta X^{-\gamma}} \tag{4.3.15}$$

This model, despite its rather peculiar appearance in having an exponent raised to an exponent, has excellent statistical properties (Ratkowsky, 1986b). One of its parameters, α, can readily be

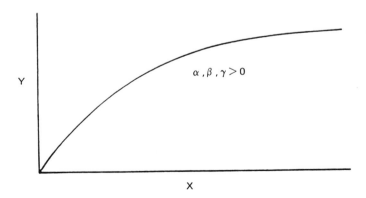

Figure 4.25

replaced by an expected-value parameter y_1, corresponding to $X = X_1$, to yield

$$(*) \quad y = y_1 X^{\beta X^{-\gamma}} / X_1^{\beta X_1^{-\gamma}} \tag{4.3.16}$$

The model is very versatile, as it can also have a maximum or a minimum and a pair of inflection points (see Sec. 6.3).

Another three-parameter model (see Figure 4.26) is that of Gunary (1970)

$$(*) \quad y = X / (\alpha + \beta X + \gamma \sqrt{X}) \tag{4.3.17}$$

This model is also versatile in being able to have a maximum or a minimum and a pair of inflection points (see Sec. 6.3).

Another three-parameter model (see Fig. 4.27) is the Holliday (1960) model,

$$(*) \quad y = 1 / (\alpha + \beta X + \gamma X^2) \tag{4.3.18}$$

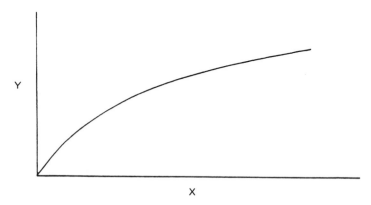

Figure 4.26

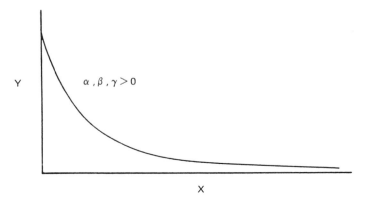

Figure 4.27

As shown by Ratkowsky (1983, Chap. 3), the statistical properties
of (4.3.18) are excellent, this model exhibiting only slightly more
nonlinearity than the corresponding model [(4.2.20)] when $\gamma = 0$.
The parameters α, β, and γ can all be replaced by expected-value
parameters y_1, y_2, and y_3, corresponding to $X = X_1$, $X = X_2$, and
$X = X_3$, respectively, by solving the three simultaneous equations

$$1/y_1 = \alpha + \beta X_1 + \gamma X_1^2$$
$$1/y_2 = \alpha + \beta X_2 + \gamma X_2^2$$
$$1/y_3 = \alpha + \beta X_3 + \gamma X_3^2$$

for α, β, and γ in terms of y_1, y_2, and y_3, but it will rarely be
necessary to use the extremely messy resulting equation because
the statistical properties of (4.3.18) are generally very good. The
form of (4.3.18) indicates that it is a generalized linear model with
linear predictor $\alpha + \beta X + \gamma X^2$ and a reciprocal link. Therefore, it
can be fitted using GLIM. A reparameterization of (4.3.18),

$$(*)\quad y = \alpha/(1 + \beta X + \gamma X^2) \tag{4.3.19}$$

has similar estimation properties to (4.3.18).

Just as (4.3.18) has been widely used in yield-density applications in agricultural research, so has the Bleasdale–Nelder (1960) model (see Figure 4.28)

$$y = (\alpha + \beta X)^{-1/\theta} \qquad (4.3.20)$$

As shown by Gillis and Ratkowsky (1978) [see also Ratkowsky, 1983, Chap. 3], the parameters α and β in (4.3.20) have extremely bad estimation properties, whereas θ is close-to-linear. The parameters α and β in (4.3.20) may be replaced by expected-value parameters y_1 and y_2, corresponding to $X = X_1$ and $X = X_2$, respectively, to yield

$$(**) \quad y = y_1 y_2 (X_2 - X_1)^{1/\theta} / [y_2^\theta (X_2 - X) + y_1^\theta (X - X_1)]^{1/\theta} \qquad (4.3.21)$$

This model has excellent estimation properties when X_1 and X_2 are chosen to be within the observed range of the data.

Another three-parameter model (see Figure 4.29) that has been proposed for use in yield-density work is that of Farazdaghi and

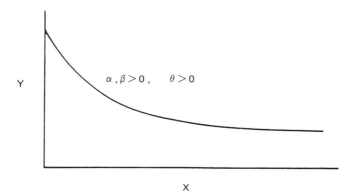

Figure 4.28

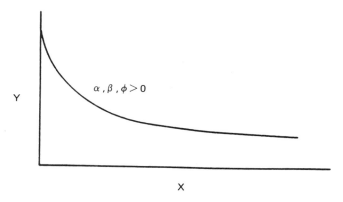

Figure 4.29

Harris (1968)

$$y = (\alpha + \beta X^\phi)^{-1} \qquad (4.3.22)$$

The parameters α and β, in common with those of α and β in the Bleasdale–Nelder model (4.3.20), have poor estimation properties, whereas ϕ is close-to-linear (Ratkowsky, 1983). The parameters α and β may be replaced by expected-value parameters y_1 and y_2, corresponding to $X = X_1$ and $X = X_2$, respectively, to yield

$$(**) \quad y = y_1 y_2 (X_2^\phi - X_1^\phi)/[y_2(X_2^\phi - X^\phi) + y_1(X^\phi - X_1^\phi)] \qquad (4.3.23)$$

This model has excellent statistical properties when X_1 and X_2 are chosen to be within the observed range of the data.

Mead (1979) proposed another solution to overcome the poor estimation properties of α and β in (4.3.20). He advocated finding an invariant value for θ in (4.3.20) over a number of data sets, in which case the precision of the estimation of α and β in (4.3.20) will be greatly improved, resulting in properties similar to those

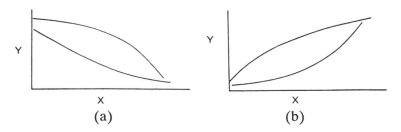

Figure 4.30

for the estimation of α and β in (4.2.20). A similar argument may be advanced for finding an invariant value of ϕ in (4.3.22).

Another widely applicable three-parameter model (see Figure 4.30a,b) is the *rational functions* or inverse polynomial model

$$(*) \quad y = (\beta + \gamma X)/(1 + \alpha X) \qquad (4.3.24)$$

Many other parameterizations of this model are possible, for example,

$$y = (1 + \gamma_1 X)/(\alpha_1 + \beta_1 X) \qquad (4.3.25)$$

$$y = (\beta_2 + \gamma_2 X)/(\alpha_2 + X) \qquad (4.3.26)$$

$$y = (\beta_3 + X)/(\alpha_3 + \gamma_3 X) \qquad (4.3.27)$$

$$y = \gamma_2 + \beta_4/(X + \alpha_2) \qquad (4.3.28)$$

where the parameters α, β, and γ are given different subscripts so that it will not be thought that they have the same meaning in each of the parameterizations. In general, (4.3.25–4.3.28) do not have as consistently good estimation properties as those of (4.3.24). Rational functions are widely used in computer applications for approximating other mathematical functions (Hastings, 1955). A rational functions model is a good starting point for empirical curve fitting when there is no theoretical basis or long-standing empirical basis for selecting an appropriate model. For rational functions models with more than three parameters, see Sections 4.4, 5.4, 5.5, and 6.5.

One note of caution: using (4.3.24), there is the possibility of obtaining a singularity in the denominator of the expression at $X = -1/\alpha$ if α is negative. This can lead to the situation depicted in Figure 4.31. This problem cannot occur when α is positive or when $X = -1/\alpha$ falls outside the observed range or range of interest of the data.

All three parameters in (4.3.24) may be replaced by expected-value parameters y_1, y_2, and y_3, corresponding to $X = X_1$, $X = X_2$, and $X = X_3$, respectively, to yield

$$(**) \quad y =$$

$$\frac{(X-X_3)(X_2-X_1)y_1y_2 + (X-X_2)(X_1-X_3)y_1y_3 + (X-X_1)(X_3-X_2)y_2y_3}{(X-X_1)(X_2-X_3)y_1 + (X-X_2)(X_3-X_1)y_2 + (X-X_3)(X_1-X_2)y_3}$$

$$(4.3.29)$$

Initial estimates for y_1, y_2, and y_3 in (4.3.29) are readily obtained, in common with other expected-value parameters, by selecting X_1,

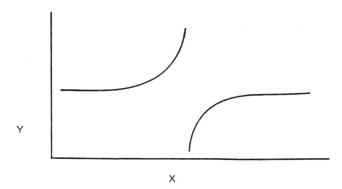

Y

X

Figure 4.31

X_2, and X_3 at the low, middle, and high end of the ordered set of Xs and reading off the "fitted" values of Y from a freehand or smooth curve drawn through the data.

Another three-parameter model (see Fig. 4.32) is the Bělehrádek (1935) equation,

$$y = \alpha(X - \beta)^\gamma \qquad\qquad (4.3.30)$$

The parameter α in (4.3.30) has poor estimation properties, whereas those of β and γ are reasonably close-to-linear. However, α may be replaced by an expected-value parameter, since (4.3.30) is of the form discussed in Section 2.5.2, to yield

$$(*) \quad y = y_1[(X - \beta)/(X_1 - \beta)]^\gamma \qquad\qquad (4.3.31)$$

where y_1 is the expected-value parameter corresponding to $X = X_1$.

Another model with three parameters (see Fig. 4.33a,b) is

$$(*) \quad y = \alpha - \beta\log(X + \gamma) \qquad\qquad (4.3.32)$$

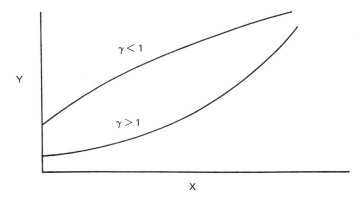

Figure 4.32

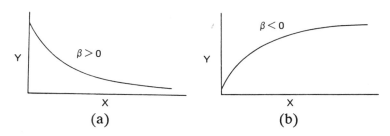

Figure 4.33

Expected-value parameters are readily found to replace α or β or both α and β. However, α, β, and γ are very similar in estimation behavior and are all sufficiently close-to-linear so as not to require replacement by expected-value parameters.

A further three-parameter model (see Fig. 4.34) is

$$(*)\quad y = \alpha \exp[\beta/(X + \gamma)] \tag{4.3.33}$$

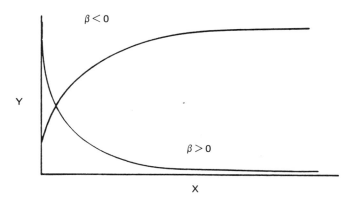

Figure 4.34

There is little difference between this parameterization and the reparameterization

$$(*) \quad y = \exp[\alpha + \beta/(X + \gamma)] \tag{4.3.34}$$

Both have reasonably close-to-linear behavior. When a multiplicative, rather than an additive, error assumption is made by taking the logarithm of both sides of (4.3.34), the resulting model becomes identical to (4.3.28), one of the parameterizations of the rational functions model, except for a change in notation.

Yet another three-parameter model (see Fig. 4.35) is

$$y = \alpha[1 - \exp(-\beta X)]^{\gamma} \tag{4.3.35}$$

often called the Chapman–Richards model and is widely used in forest research. Frequently, its worst-behaving parameter is α, especially when there are few data points near the asymptote. This parameter can be replaced by an expected-value parameter y_1,

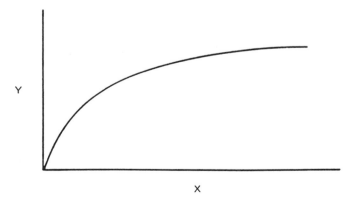

Figure 4.35

corresponding to $X = X_1$, to give

$$(*) \quad y = y_1 \{[1 - \exp(-\beta X)]/[1 - \exp(-\beta X_1)]\}^\gamma \tag{4.3.36}$$

There are other circumstances, however, when the worst-behaving parameter of (4.3.35) is not α but either β or γ. If β is the parameter with the most nonlinear behavior, it can be replaced by an expected-value parameter y_2, corresponding to $X = X_2$, to give

$$(*) \quad y = \alpha \{1 - [1 - (y_2/\alpha)^{1/\gamma}]^{X/X_2}\}^\gamma \tag{4.3.37}$$

Similarly, if γ is the worst-behaving parameter, one can replace it with y_3, corresponding to $X = X_3$, yielding

$$(*) \quad y = \alpha(y_3/\alpha)^{\log[1-\exp(-\beta X)]/\log[1-\exp(-\beta X_3)]} \tag{4.3.38}$$

This model, in any of the forms given by (4.3.35–4.3.38), is also capable of having an inflection point within the range of interest of the data (see Sec. 5.3 and 5.4).

Another three-parameter model (see Fig. 4.36) is

$$(*) \quad y = \exp(\alpha + \beta X + \gamma X^2) \tag{4.3.39}$$

This model has excellent statistical properties, with all its parameters having similar statistical behavior. It is a generalized linear model, with linear predictor $\alpha + \beta X + \gamma X^2$ and a logarithmic link function. It can therefore be fitted using GLIM. This model is also capable of having a pair of inflection points and a maximum or a minimum (see Sec. 6.3).

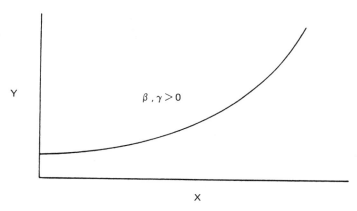

Figure 4.36

4.4   FOUR-PARAMETER CURVES

The two-surface Langmuir model (see Fig. 4.37)

$$y = \alpha\beta X/(1+\alpha X) + \gamma\delta X/(1+\gamma X) \qquad\qquad (4.4.1)$$

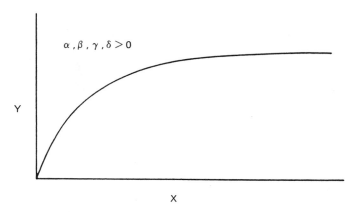

Figure 4.37

exhibited very poor statistical behavior when it was used for data on the sorption of phosphate by soil and on the inhibition of histidine uptake (Ratkowsky, 1986b). The reparameterization

$$y = X/(\theta_1 + \theta_2 X) + X/(\theta_3 + \theta_4 X) \tag{4.4.2}$$

is no better in estimation behavior than (4.4.1).

Another four-parameter model (see Figure 4.38a,b) is the rational functions model

$$y = (\beta + \gamma X)/(1 + \alpha X + \delta X^2) \tag{4.4.3}$$

Model (4.4.3) represents the next level of complexity over (4.3.24) or its reparameterizations (4.3.25–4.3.28). Many other reparameterizations of (4.4.3) are possible, but (4.4.3) seems best for general use. When (4.4.3) is used to fit convex/concave functions, as in this chapter, there is a danger of overparameterization (see Ratkowsky, 1987). This model can also have inflection points (see Sec. 5.4) and maxima or minima (see Sec. 6.4).

A very widely used and important model (see Figure 4.39) is the classical sum of exponentials model

$$y = \alpha \exp(-\beta X) + \gamma \exp(-\delta X) \tag{4.4.4}$$

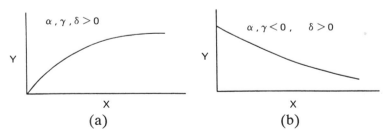

(a)   (b)

Figure 4.38

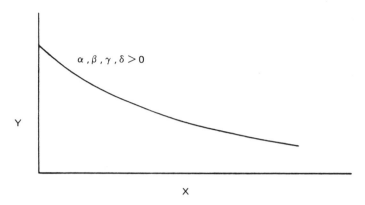

Figure 4.39

which finds application in many different types of compartment modeling. Unfortunately, the parameterization given by (4.4.4) is often far-from-linear, in which case convergence is extremely difficult and the estimation of the parameters may be extremely biased.

A slight improvement in statistical behavior over (4.4.4) may be achieved with

$$y = \alpha \lambda_1^X + \gamma \lambda_2^X \qquad (4.4.5)$$

where $\exp(-\beta)$ has been replaced by λ_1 and $\exp(-\delta)$ by λ_2. The improvement in estimation properties is slight because it is the parameters α and γ that are generally the ones having the worst estimation properties. However, these may be replaced by expected-value parameters to give

$$y = \frac{y_1(\lambda_1^X \lambda_2^{X_2} - \lambda_2^X \lambda_1^{X_2}) - y_2(\lambda_1^X \lambda_2^{X_1} - \lambda_2^X \lambda_1^{X_1})}{\lambda_1^{X_1} \lambda_2^{X_2} - \lambda_1^{X_2} \lambda_2^{X_1}} \qquad (4.4.6)$$

As always with expected-value parameters, the behavior of y_1 and y_2 in estimation is excellent. Since the behavior of λ_1 and λ_2 is reasonably good, (4.4.6) may be a suitable model for use where theory indicates that the sum of two exponentials is an appropriate model.

Another four-parameter model (see Fig. 4.40) is the generalized hyperbola (Campbell and Keay, 1970)

$$y = \alpha - \beta/(1 + \gamma X)^{1/\delta} \qquad (4.4.7)$$

The estimation behavior of this model is poor. A general principle is that a model with more than three parameters is rarely a parsimonious model for fitting a convex or concave curve. Therefore, one should use a model with as few parameters as possible unless theory dictates a more complicated model *and* the data are sufficiently good so that the residual variance of the data around the fitted model is small. With empirical modeling, extra parameters should be added *only* when goodness-of-fit tests tell the user that the simpler models are not fitting well. A variant of (4.4.7) is discussed in Chapter 8 as (8.1.15).

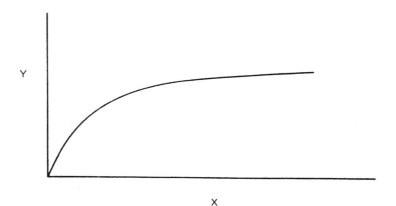

Y

X

Figure 4.40

Another four-parameter model (see Fig. 4.41) is that due to Prunty (1983),

$$y = \alpha[1 + \beta(X + \gamma) - \{1 + [\beta(X + \gamma)]^{\delta}\}^{1/\delta}] \qquad (4.4.8)$$

Model (4.4.8) is a smooth function made up of two straight-line segments as a limiting case when $\delta \to \infty$. The statistical properties in estimation of (4.4.8) are poor and, when this model was fitted to a wide variety of data on crop growth, it was found (Ratkowsky, unpublished results) that it did not fit data any better than a number of more traditional models, despite the fact that (4.4.8) has four parameters. In fact, the goodness-of-fit of (4.4.8) is very similar to that of (4.4.7), but neither is consistently better than various three-parameter models with good statistical properties.

Another four-parameter model (see Figure 4.42a,b), which is made up of two straight-line segments as a limiting case, like (4.4.8), is

$$(*) \quad y = \alpha + \beta X - \{(\gamma + \beta X)^2 + \delta^2\}^{1/2} \qquad (4.4.9)$$

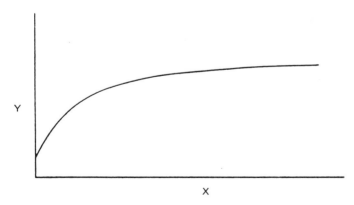

Figure 4.41

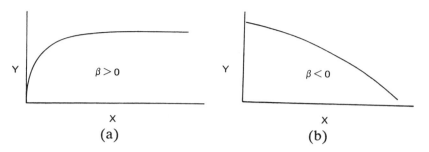

Figure 4.42

Model (4.4.9) is a simplified "bent-hyperbola" model, which reduces to two intersecting straight lines when $\delta = 0$. For $\delta > 0$, the presence of this term allows for a smooth transition between the intersecting straight lines. Good fits using (4.4.9) and good statistical properties were obtained for data on electrode potential versus ion concentration (data courtesy of John MacFarlane, Hawkesbury Agricultural College) and also for Data Sets 2 and 3, Table 6.18, of Ratkowsky (1983).

4.5   FIVE-PARAMETER AND SIX-PARAMETER CURVES

The model (see Fig. 4.43)

$$y = \alpha \exp(-\beta X) + \gamma \exp(-\delta X) + \eta \exp(-\nu X) \tag{4.5.1}$$

is an extension of (4.4.4), the classic sum of exponentials model. It is even more difficult to achieve convergence with (4.5.1) than with (4.4.4). As with the latter model, a slight improvement in the estimation behavior may be achieved by replacing $\exp(-\beta)$,

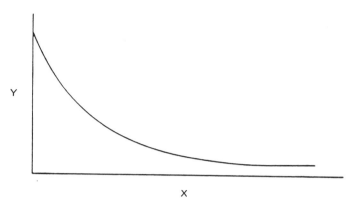

Figure 4.43

$\exp(-\delta)$, and $\exp(-\gamma)$ by λ_1, λ_2, and λ_3, respectively, to yield

$$y = \alpha\lambda_1^X + \gamma\lambda_2^X + \eta\lambda_3^X \qquad (4.5.2)$$

However, the improvement in statistical behavior over (4.5.1) is slight because the parameters exhibiting the poorest statistical properties are generally α, γ, and η (although not necessarily always). These may be replaced by expected-value parameters, which have excellent estimation properties. As the algebra is extremely messy, it is best to express the model in matrix form. Thus, one may write (4.5.2) as

$$y = (\lambda_1^X \; \lambda_2^X \; \lambda_3^X) \begin{bmatrix} \alpha \\ \gamma \\ \eta \end{bmatrix}$$

One now chooses three values of X, calling them X_1, X_2, and X_3, which may be the lowest, midmost, and highest X values in the data set, although they may be any three widely spaced X values within the observed data range. Use of $X_1, X_2,$ and X_3 gives the following expected-value parameters:

$$\begin{bmatrix} y_1 \\ y_2 \\ y_3 \end{bmatrix} = \begin{bmatrix} \lambda_1^{X_1} & \lambda_2^{X_1} & \lambda_3^{X_1} \\ \lambda_1^{X_2} & \lambda_2^{X_2} & \lambda_3^{X_2} \\ \lambda_1^{X_3} & \lambda_2^{X_3} & \lambda_3^{X_3} \end{bmatrix} \begin{bmatrix} \alpha \\ \gamma \\ \eta \end{bmatrix}$$

The "original" parameters α, γ, and η can be eliminated using matrix inversion to produce the following expression:

$$y = \begin{pmatrix} \lambda_1^X & \lambda_2^X & \lambda_3^X \end{pmatrix} \begin{bmatrix} \lambda_1^{X_1} & \lambda_2^{X_1} & \lambda_3^{X_1} \\ \lambda_1^{X_2} & \lambda_2^{X_2} & \lambda_3^{X_2} \\ \lambda_1^{X_3} & \lambda_2^{X_3} & \lambda_3^{X_3} \end{bmatrix}^{-1} \begin{bmatrix} y_1 \\ y_2 \\ y_3 \end{bmatrix} \qquad (4.5.3)$$

Model (4.5.3) contains six parameters, $\lambda_1, \lambda_2, \lambda_3, y_1, y_2,$ and y_3, the last three being expected-value parameters. These have excellent estimation properties and, when those of $\lambda_1, \lambda_2,$ and λ_3 are reasonably good, (4.5.3) is a suitable model for use where theory indicates, as in compartment modeling, that the sum of three exponentials is an appropriate model.

Model (4.5.3) also includes a reparameterization of the following five-parameter model:

$$y = \alpha \lambda_1^X + \gamma \lambda_2^X + \eta \qquad (4.5.4)$$

One simply replaces λ_3 in (4.5.3) with 1 to give

$$y = \begin{pmatrix} \lambda_1^X & \lambda_2^X & 1 \end{pmatrix} \begin{bmatrix} \lambda_1^{X_1} & \lambda_2^{X_1} & 1 \\ \lambda_1^{X_2} & \lambda_2^{X_2} & 1 \\ \lambda_1^{X_3} & \lambda_2^{X_3} & 1 \end{bmatrix}^{-1} \begin{bmatrix} y_1 \\ y_2 \\ y_3 \end{bmatrix} \qquad (4.5.5)$$

Another five-parameter model (see Figure 4.44a–d), made up of two intersecting straight lines as a limiting case, is

$$y = \alpha + \beta(X - \delta) - \gamma[(X - \delta)^2 + \epsilon^2]^{1/2} \qquad (4.5.6)$$

Model (4.5.6) is a reparameterization of one proposed by Griffiths and Miller (1973) [see Ratkowsky, 1983, model (6.12)]. The worst-behaving parameter in (4.5.6) is often α, followed by γ. Both of these parameters can be replaced by expected-value parameters, but the resulting expression is messy. Replacing α alone by an expected-value parameter y_1, corresponding to $X = X_1$, yields

$$y = y_1 + \beta(X - X_1) - \gamma\{[(X - \delta)^2 + \epsilon^2]^{1/2} - [(X_1 - \delta)^2 + \epsilon^2]^{1/2}\} \qquad (4.5.7)$$

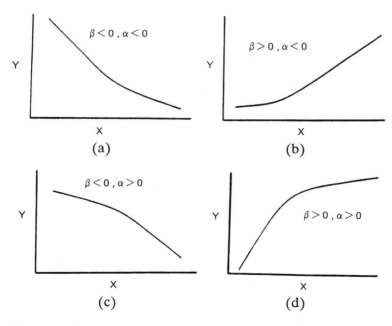

Figure 4.44

which is sometimes reasonably close-to-linear in behavior.

An alternative five-parameter model (see Figure 4.45a–d) which, like (4.5.6) and (4.5.7), approaches two straight lines as a limiting case, is

$$y = \alpha + \beta(X - \delta) - \gamma(X - \delta)\tanh[(X - \delta)/\epsilon] \qquad (4.5.8)$$

Model (4.5.8) was proposed by Bacon and Watts (1971) and, like (4.5.6), (4.5.7), and the four-parameter model (4.4.9), it has a smooth transition between segments that approach a straight line. Although this smooth transition is achieved by use of the hyperbolic tangent in (4.5.8), Bacon and Watts (1971) considered other transition functions but found that the differences between the functions were all extremely small.

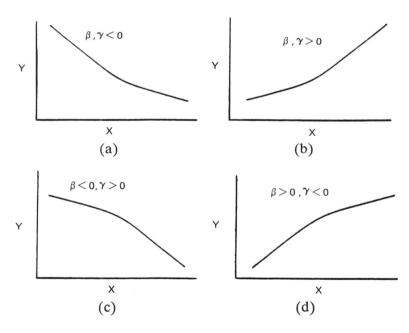

Figure 4.45

The worst-behaving parameter in (4.5.8) is often γ, which may readily be replaced by an expected-value parameter y_1, corresponding to $X = X_1$, to yield

$$y = y_1 + \beta(X - X_1) - \gamma\{(X - \delta)\tanh[(X - \delta)/\epsilon]$$
$$- (X_1 - \delta)\tanh[(X - \delta)/\epsilon]\} \quad (4.5.9)$$

which is sometimes reasonably close-to-linear in estimation.

5

Models with One X Variable, Sigmoidally Shaped Curves

This chapter concerns models with a single explanatory variable in which the graph of Y versus X has an inflection point but lacks a maximum or a minimum, except possibly at $\pm\infty$ or at infeasible values of X. Thus, the typical shape of curves in this chapter is sigmoidal.

5.1 ONE-PARAMETER CURVES

A model having a sigmoidal shape with but a single parameter (see Fig. 5.1) is

$$(*) \quad y = 1 - \exp(-X^{\alpha}) \tag{5.1.1}$$

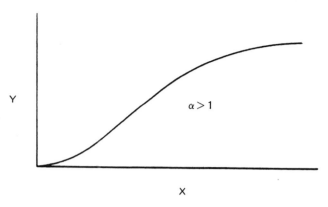

Y

$\alpha > 1$

X

Figure 5.1

The inflection point occurs at $X = [(\alpha - 1)/\alpha]^{1/\alpha}$. The statistical properties in estimation of this model are reasonably good, but the range of the response variable Y is restricted to lie between zero and one. Replacing α by an expected-value parameter y_1, corresponding to $X = X_1$, results in

$$(**) \quad y = 1 - \exp\{-[-\log(1-y_1)]\}^{\log X/\log X_1} \qquad (5.1.2)$$

The model

$$(*) \quad y = \exp(-X^\alpha) \qquad (5.1.3)$$

has identical estimation properties to the above, the only difference being that y decreases with increasing values of X instead of increasing. Replacing α by an expected-value parameter y_1, corresponding to $X = X_1$, results in

$$(**) \quad y = \exp\{-[\log(1/y_1)]^{\log X/\log X_1} \qquad (5.1.4)$$

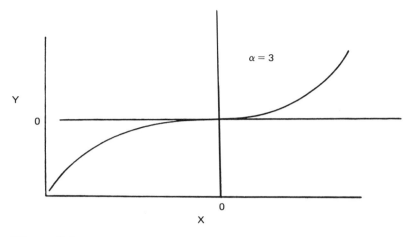

Figure 5.2

Other one-parameter models with the capability of having an inflection point exist, but tend to be too restrictive for general use. For example, consider (4.1.7) of Chapter 4,

$$y = X^\alpha$$

When α is an odd integer of absolute magnitude greater than 2, an inflection point occurs at $X = 0$, as in Fig. 5.2. However, when α is not exactly an integer, the term X^α cannot be evaluated for negative X. Hence, the full curve does not appear to have a place in nonlinear regression modeling, where the best-fitting value of the parameter α is usually determined by the criterion of least squares.

5.2 TWO-PARAMETER CURVES

A model with two parameters (see Figure 5.3) having an inflection point is

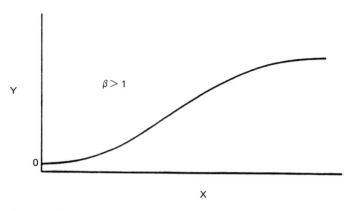

Figure 5.3

$$(*) \quad y = 1 - \exp(-\alpha X^{\beta}) \tag{5.2.1}$$

The inflection point occurs at $X = [(\beta - 1)/\alpha\beta]^{1/\beta}$ and is thus more versatile than (5.1.1). Similarly, the model $y = \exp(-\alpha X^{\beta})$ has identical estimation properties to (5.2.1), with y decreasing rather than increasing as X increases.

Another two-parameter model (see Fig. 5.4) is

$$y = 1 - \alpha \exp(-X^{\beta}) \tag{5.2.2}$$

The inflection point occurs at $X = [(\beta - 1)/\beta]^{1/\beta}$. The estimation properties of (5.2.2) are not as good as those of (5.2.1), a different basic model with a similar shape, the behavior of the estimator of β tending to be inferior to that of the estimator of β in (5.2.1). Although β in (5.2.2) can be replaced by an expected-value parameter, the resulting expression is messy. The model $y = \alpha \exp(-X^{\beta})$ is similar, except that y decreases as X increases.

Another two-parameter model with an inflection point (see Fig. 5.5) is

$$y = 1 - \exp(-\alpha \beta^{X}) \tag{5.2.3}$$

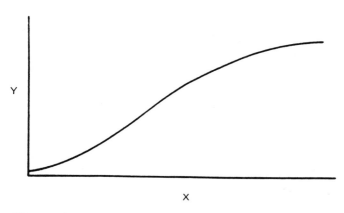

Figure 5.4

Of the two parameters, α tends to have the poorer estimation behavior. An alternative parameterization to (5.2.3) is

$$(*) \quad y = 1 - \exp[-\exp(\gamma - \delta X)] \tag{5.2.4}$$

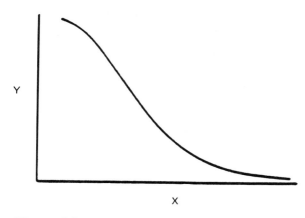

Figure 5.5

The statistical properties of (5.2.4) are quite good, the estimators of the parameters γ and δ exhibiting a similar amount of
nonlinearity. Another alternative is to replace α in (5.2.3) by an
expected-value parameter y_1, evaluated at $X = X_1$, to yield

$$(*) \quad y = 1 - (1 - y_1)^{\exp[-\delta(X - X_1)]} \tag{5.2.5}$$

in which δ in (5.2.5) is identical to δ in (5.2.4). Similar model
functions to the above, in which y increases with increasing X
instead of decreasing, are obtained by deleting the first term and
following minus sign of the right-hand sides of (5.2.3–5.2.5). Thus,
for example, one would obtain

$$(*) \quad y = \exp[-\exp(\gamma - \delta X)] \tag{5.2.6}$$

from (5.2.4).

5.3 THREE-PARAMETER CURVES

One of the most versatile and useful models (see Figure 5.6) for
fitting sigmoidal responses having a lower asymptote of zero and a
finite upper asymptote is the logistic model

$$(*) \quad y = \alpha/[1 + \exp(\beta - \gamma X)] \tag{5.3.1}$$

The curve is skew-symmetric, with an inflection point at $X = \beta/\gamma$,
$Y = \alpha/2$. The parameterization given by

$$(*) \quad y = \alpha/\{1 + \exp[-\gamma(X - \delta)]\} \tag{5.3.2}$$

is even slightly better than (5.3.1), as the estimation properties of
δ in (5.3.2) tends to be better than the estimation properties of β

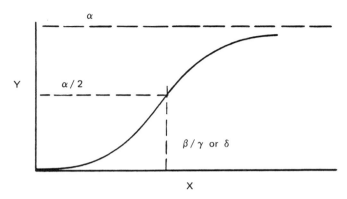

Figure 5.6

in (5.3.1). There is little to choose between the two parameterizations. Information on other parameterizations can be found in Ratkowsky (1983), Section 4.4.2. Some of these parameterizations will be referred to again in Section 5.4, which deals with the four-parameter logistic, an extension of (5.3.1) or (5.3.2) that allows for a lower asymptote that is different from zero. Reversing the signs of both β and γ allows y to decrease with increasing X.

Although the symmetry of this model, which makes the upper position of the curve a mirror image of the lower portion, seems to be a restriction, the model has found numerous uses in both theoretical and empirical modeling.

Another versatile and useful three-parameter curve (see Figure 5.7) with an inflection point is the Gompertz model,

$$(*) \quad y = \alpha \exp[-\exp(\beta - \gamma X)] \tag{5.3.3}$$

In common with the logistic model (5.3.1) or (5.3.2), (5.3.3) has asymptotes at $y = 0$ and $y = \alpha$ but, unlike the logistic model, the Gompertz is asymmetric about its inflection point which, like (5.3.1), also occurs at β/γ. Other parameterizations of (5.3.3)

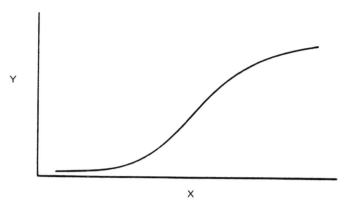

Figure 5.7

include

$$(*) \quad y = \alpha \exp\{-\exp[-\gamma(X - \epsilon)]\} \tag{5.3.4}$$

and

$$(*) \quad y = \exp(\eta - \nu\theta^X) \tag{5.3.5}$$

but there is little to choose between these model functions. All have similar estimation properties.

In the parameterization given by (5.3.5), it is easy to see that, if one uses a multiplicative error assumption (equivalent to taking the logarithm of both sides of the equation), the response becomes identical to the asymptotic regression model (4.3.1–4.3.9).

A further three-parameter curve (see Figure 5.8) with an inflection point but without maxima or minima is one that appeared in Section 4.3 as (4.3.33),

$$(*) \quad y = \alpha \exp[\beta/(X + \gamma)] \tag{5.3.6}$$

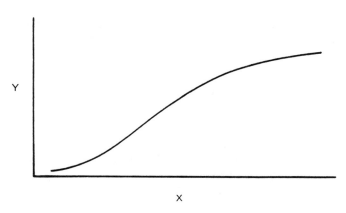

Figure 5.8

The inflection point occurs at $X = -\beta/2 - \gamma$.

Another three-parameter curve (see Figure 5.9) with the capability of having an inflection point is one that appeared in Section 4.3 as (4.3.35),

$$y = \alpha[1 - \exp(-\beta X)]^{\gamma} \tag{5.3.7}$$

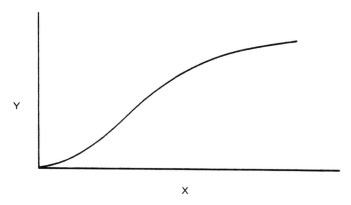

Figure 5.9

As noted in Section 4.3, α is often the worst-behaving parameter, but it may be replaced by an expected-value parameter y_1, corresponding to $X = X_1$, to yield

$$(*) \quad y = y_1[1 - \exp(-\beta X)]^\gamma / [1 - \exp(-\beta X_1)]^\gamma \qquad (5.3.8)$$

which already appeared in Section 4.3 as (4.3.36). Sometimes it is β or γ that exhibits the greatest amount of nonlinear behavior rather than α. Under those circumstances, one should use the model functions (4.3.37) or (4.3.38), respectively.

Yet another three-parameter curve (see Figure 5.10) with an inflection point but no maxima or minima, except at $X = 0$, the boundary of permissible observed values of X, is the extended Langmuir model, in Section 4.3 as (4.3.10),

$$y = \alpha\beta X^{1-\gamma}/(1 + \beta X^{1-\gamma}) \qquad (5.3.9)$$

As noted in Section 4.3, γ is often a close-to-linear parameter, but α and β may both be far-from-linear. The reparameterization

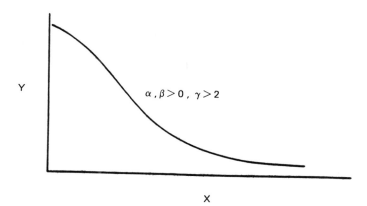

Figure 5.10

given as (4.3.11),

$$y = 1/(\theta_1 + \theta_2 X^{\gamma-1})$$

is better behaved but should be used with caution.

Another three-parameter curve (see Figure 5.11a,b) with an inflection point is

$$y = \alpha - \beta \exp(-X^\gamma) \tag{5.3.10}$$

The inflection point occurs at $X = [(\gamma - 1)/\gamma]$. The worst-behaving parameter estimator is usually β, which can be replaced by an expected-value parameter y_1, corresponding to $X = X_1$, to yield

$$(*) \quad y = \alpha - (\alpha - y_1)\exp(X_1^\gamma - X^\gamma) \tag{5.3.11}$$

Another three-parameter curve (see Fig. 5.12) with an inflection point is

$$y = \alpha[1 - \exp(-\beta X^\gamma)] \tag{5.3.12}$$

the inflection point occurring at $X = [(\gamma - 1)/\beta\gamma]^{1/\gamma}$ with $\gamma > 1$. When β is the worst-behaving parameter, as often happens when

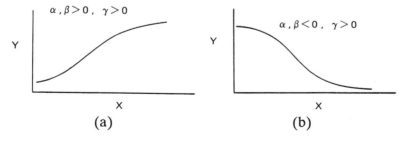

(a)   (b)

Figure 5.11

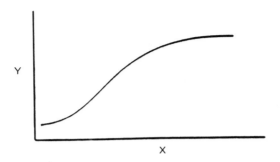

Figure 5.12

$\gamma > 3$, it can be replaced by an expected-value parameter y_1, corresponding to $X = X_1$, to yield

$$y = \alpha\{1 - [1 - (y_1/\alpha)]^{(X/X_1)^\gamma}\} \qquad (5.3.13)$$

which may then be reasonably close-to-linear. When α is the worst-behaving parameter, as often happens when $\gamma < 2$, it can be replaced by an expected-value parameter y_2 (say), corresponding to $X = X_2$, to yield

$$y = y_2[1 - \exp(-\beta X^\gamma)]/[1 - \exp(-\beta X_2^\gamma)] \qquad (5.3.14)$$

This latter idea may be combined with replacing β by $-\log\delta$, which gives

$$y = y_2[1 - \delta^{X^\gamma}]/[1 - \delta^{X_2^\gamma}] \qquad (5.3.15)$$

and which may be reasonably close-to-linear in estimation.

A model that was considered in Chapter 3 for relating moisture content in foodstuffs to water activity was the GAB model (3.2.11)

(see Figure 5.13)

$$y = Y_M CkX/[(1-kX)(1-kX+CkX)]$$

The estimation properties of the parameters in this parameterization are poor, but if one divides through by $Y_M Ck$, using
the parameters-in-denominator principle of reparameterization
(Sec. 2.3.2) and then multiplies out terms in the denominator, one
obtains

$$(*) \quad y = X/(\alpha + \beta X - \gamma X^2) \tag{5.3.16}$$

which is really a rearranged form of the Hailwood-Horrobin (1946)
model (3.2.12). The statistical properties in estimation of (5.3.16)
tend to be very good.

Another three-parameter curve (see Fig. 5.14) with an inflection
point is

$$y = \alpha - \beta \log(X^{-\gamma} - 1) \tag{5.3.17}$$

This model proved to have reasonably good estimation properties
when tested on several sets of soil data. The inflection point

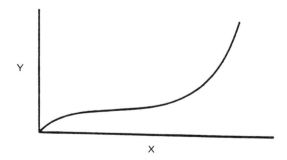

Figure 5.13

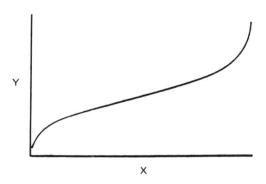

Figure 5.14

occurs at $X = [1/(1 + \gamma)]^{1/\gamma}$. The worst-behaving parameter was γ for some of the data sets and α for some of the other data sets. Replacing γ in (5.3.17) with an expected-value parameter y_1, corresponding to $X = X_1$, yields

$$(*) \quad y = \alpha - \beta \log\{[1 + \exp((\alpha - y_1)/\beta)]^{\log X/\log X_1} - 1\} \qquad (5.3.18)$$

a rather messy expression, whereas replacing α in (5.3.17) with an expected-value parameter y_2, corresponding to $X = X_2$, yields

$$(*) \quad y = y_2 - \beta \log[(X^{-\gamma} - 1)/(X_2^{-\gamma} - 1)]. \qquad (5.3.19)$$

Both (5.3.18) and (5.3.19) would be closer-to-linear in behavior than (5.3.17), the choice of which one to use depending on whether it is γ or α, respectively, that is mainly responsible for the nonlinearity in (5.3.17).

5.4 FOUR-PARAMETER CURVES

The simple logistic model (5.3.1) is readily extended to allow a nonzero lower asymptote δ,

$$(*) \quad y = \delta + \alpha/[1 + \exp(\beta - \gamma X)] \tag{5.4.1}$$

As indicated in Figure 5.15, the upper asymptote (Max) is $\alpha + \delta$. The nonlinear behavior of (5.4.1) is only slightly greater than that of (5.3.1), as a result of adding the extra parameter. Ratkowsky and Reedy (1986) studied (5.4.1) and a variety of its reparameterizations in combination with data on radioligand and other assays. Generally, for assay work, the response Y is logistic in $\log X$, rather than X itself. Thus, one would use the modification of (5.4.1) where X is replaced by $\log X$. It was found (Ratkowsky and Reedy, 1986) that five parameterizations had good statistical properties, these being

$$(*) \quad y = \text{Min} + \text{Range}/[1 + \exp(\beta - \gamma \log X)] \tag{5.4.2}$$

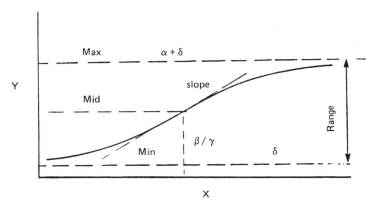

Figure 5.15

$$(*) \quad y = \text{Min} + 4(\text{Slope})/\{\gamma[1 + \exp(\beta - \gamma \log X)]\} \qquad (5.4.3)$$

$$(*) \quad y = \text{Min} + (\text{Max} - \text{Min})/\{1 + \exp[\gamma(\beta - \log X)]\} \qquad (5.4.4)$$

$$(*) \quad y = \text{Mid} - (\text{Range}/2)\tanh[(\gamma/2)(\beta - \log X)] \qquad (5.4.5)$$

$$(*) \quad y = \text{Min} + (\text{Max} - \text{Min})/[1 + (\beta/X)^{\gamma}] \qquad (5.4.6)$$

where Min is the lower asymptote, Max is the upper asymptote, Range = Max − Min (that is, the range of the fitted response), Mid = (Min + Max)/2 (that is, the middle response), Slope is the slope of the response at the inflection point, and γ (but not β) has a consistent meaning throughout (5.4.2–5.4.6). For use of these parameterizations when the response is logistic in X rather than $\log X$, one simply replaces $\log X$ in (5.4.2–5.4.6) by X in each of those equations. [For (5.4.6), this means replacing X with $\exp(X)$.]

Two further reparameterizations of (5.4.1) were studied by Ratkowsky and Reedy (1986), but these are not recommended because of their poor estimation properties. They are

$$y = \text{Min} + \text{Range}/[1 + \beta/X^{\gamma}] \qquad (5.4.7)$$

and

$$y = \text{Max} - \text{Range} + \text{Range}/[1 + (1/\beta X^{\gamma})] \qquad (5.4.8)$$

Any of (5.4.1–5.4.6) can be reduced to the three-parameter logistic by setting Min = 0 (or $\delta = 0$). [For (5.4.5), this means replacing Mid by Max/2 and Range by Max.] The four-parameter logistic shares with the three-parameter logistic the property of being symmetric about its inflection point. The logistic can also have the shape of a decreasing sigmoidal curve (see Figure 5.16) by reversing the signs of β and γ in models (5.4.1–5.4.5) and (5.3.1).

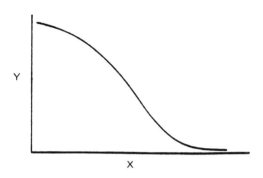

Figure 5.16

A four-parameter model (see Figure 5.17a,b) suitable for data that are asymmetric about the inflection point is the four-parameter Gompertz, an extension of (5.3.3),

$$(*) \quad y = \delta + \alpha \exp[-\exp(\beta - \gamma X)] \qquad\qquad (5.4.9)$$

As noted in Section 5.3, certain other parameterizations of (5.3.3) had similar statistical properties. Similarly, the reparameterizations

$$(*) \quad y = \delta + \alpha \exp\{-\exp[\gamma(X - \epsilon)]\} \qquad\qquad (5.4.10)$$

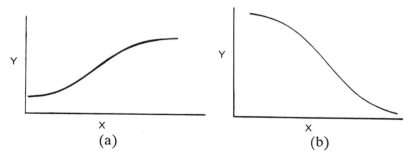

(a)   (b)

Figure 5.17

and

$$(*) \quad y = \delta + \exp(\eta - \nu \theta^X) \qquad\qquad (5.4.11)$$

may be used in place of (5.4.9). All three parameterizations, (5.4.9–5.4.11), have good estimation properties.

Another four-parameter model (see Figure 5.18) that has found wide use in applied science because of the belief on the part of users that it is a very "flexible" curve is the Richards (1959) model

$$y = \alpha/[1 + \exp(\beta - \gamma X)]^{1/\delta} \qquad\qquad (5.4.12)$$

This model includes the three-parameter logistic model (5.3.1) as a special case ($\delta = 1$) and the three-parameter Gompertz model (5.3.3–5.3.5) as a special case ($\delta \to 0$). It has, nevertheless, a lower asymptote of zero, which makes it less flexible than the four-parameter logistic (5.4.1) or the four-parameter Gompertz (5.4.9–5.4.11). Unfortunately, the Richards model has *very poor* statistical properties in estimation, possessing considerable intrinsic nonlinearity (Ratkowsky, 1983, Chap. 4) and generally having 2

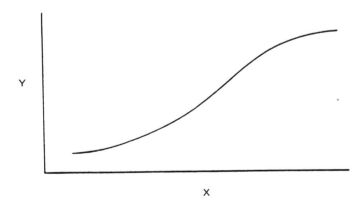

Y

X

Figure 5.18

or 3 parameters whose estimators have poor statistical properties. Because of the high intrinsic nonlinearity of this model, there is little incentive to search for possible reparameterizations of (5.4.12). The Richards model (5.4.12) exhibits more undesirable nonlinear regression behavior than almost any nonlinear regression model in common use. The continued use of this model is not recommended.

Another four-parameter sigmoidal curve (see Figure 5.19a,b) is the Morgan, Mercer, and Flodin (1975) model

$$y = (\beta\gamma + \alpha X^\delta)/(\gamma + X^\delta) \qquad (5.4.13)$$

The worst-behaving parameter is γ (Ratkowsky, 1983, Chap. 4), which can be replaced by $\exp(\gamma)$ to give

$$(*)\quad y = [\beta\exp(\gamma) + \alpha X^\delta]/[\exp(\gamma) + X^\delta] \qquad (5.4.14)$$

The behavior in estimation of this reparameterization is reasonably good. An even better solution, however, is to replace γ by an expected-value parameter y_1, corresponding to $X = X_1$, to give

$$(*)\quad y = [\beta(\alpha - y_1)X_1^\delta + \alpha(y_1 - \beta)X^\delta]/[(\alpha - y_1)X_1^\delta + (y_1 - \beta)X^\delta] \qquad (5.4.15)$$

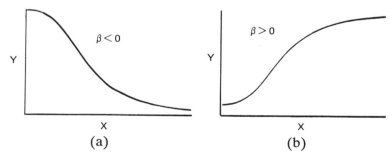

(a)   (b)

Figure 5.19

This parameterization is reasonably close-to-linear in estimation.

Another four-parameter model (see Figure 5.20a,b) with an inflection point is the Weibull-type model (Yang et al., 1978; Reid, 1978)

$$y = \alpha - \beta \exp(-\gamma X^\delta) \tag{5.4.16}$$

As shown by Ratkowsky (1983, Chap. 4), the parameter with the worst estimation behavior is γ, which can be replaced by $\exp(-\gamma)$ to give

$$(*) \quad y = \alpha - \beta \exp[-\exp(-\gamma)X^\delta] \tag{5.4.17}$$

Although the behavior in estimation of this parameterization is reasonably close-to-linear, an alternative approach is to replace γ with an expected-value parameter y_1 corresponding to $X = X_1$ to give

$$(*) \quad y = \alpha - \beta[(\alpha - y_1)/\beta]^{(X/X_1)^\delta} \tag{5.4.18}$$

For various data sets, I have found (unpublished results) that, although the estimation behavior of δ is always excellent, the

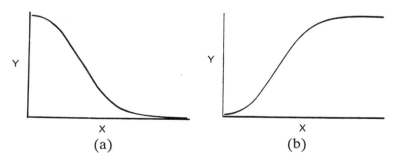

(a)   (b)

Figure 5.20

behavior of α and β (as well as that of γ) in (5.4.16) is not. However, α and β in (5.4.16) may be replaced by expected-value parameters in the same way that α and β in (4.3.1) was replaced by Schnute and Fournier (1980) and, at the same time, γ may be replaced by a parameter k, which has better estimation properties than those of γ. The resulting parameterization has the same form as (4.3.8),

$$(*) \quad y = y_1 + (y_2 - y_1)(1 - k^{m-1})/(1 - k^{n-1}) \tag{5.4.19}$$

where m is now given by

$$m = 1 + (n - 1)(X^\delta - X_1^\delta)/(X_2^\delta - X_1^\delta)$$

with y_1 and y_2 being expected-value parameters corresponding to $X = X_1$ and $X = X_2$, respectively. The resulting model (5.4.19) is formally the same as (4.3.8), but with X, X_1, and X_2 replaced by X^δ, X_1^δ, and X_2^δ, respectively. The estimation properties of the four parameters y_1, y_2, k, and δ are excellent. As with (4.3.8), y_1 and y_2 in (5.4.19) correspond to prechosen values X_1 and X_2 of the explanatory variable X. These may be the lowest and highest observed values of X in the data set, but this is not a requirement, provided that they are within the observed range of the data. Model (5.4.19) should find widespread use in fitting ascending and descending sigmoidally shaped curves.

Another four-parameter sigmoidal model (see Figure 5.21) is the extension of the Chapman-Richards model (4.3.35) to allow for a nonzero lower asymptote when $X = 0$,

$$y = \alpha[1 - \exp(-\beta X)]^\gamma + \delta \tag{5.4.20}$$

As was the case with (4.3.35), the worst-behaving parameter is often α, which can be replaced by an expected-value parameter y_1,

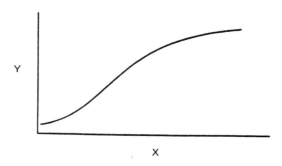

Figure 5.21

corresponding to $X = X_1$, to give

$$(*) \quad y = \delta + (y_1 - \delta)[1 - \exp(-\beta X)]^{\gamma}/[1 - \exp(-\beta X_1)]^{\gamma} \quad (5.4.21)$$

This model must be used with caution, as there are situations in which the worst-behaving parameter of (5.4.20) is not α but one of the other parameters. If β is the parameter with the most nonlinear behavior, it can be replaced by an expected-value parameter y_2, corresponding to $X = X_2$, to give

$$(*) \quad y = \delta + \alpha\{1 - (1 - [(y_2 - \delta)/\alpha]^{1/\gamma})^{X/X_2}\}^{\gamma} \quad (5.4.22)$$

Similarly, if γ is the worst-behaving parameter, one can replace it with y_3, corresponding to $X = X_3$, yielding

$$(*) \quad y = \delta + \alpha[(y_3 - \delta)/\alpha]^{\log[1-\exp(-\beta X)]/\log[1-\exp(-\beta X_3)]} \quad (5.4.23)$$

Another four-parameter model (see Fig. 5.22) with a sigmoidal shape is

$$y = \alpha/[1 + \beta(X - \gamma)^{\delta}] \quad (5.4.24)$$

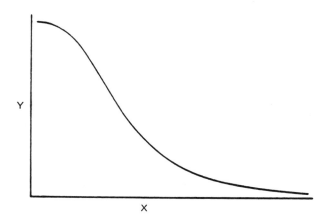

Figure 5.22

for which $X \geq \gamma$. This model tends to be rather unstable, with two or more of its parameters exhibiting considerable nonlinear behavior. Its use is not recommended.

A further four-parameter sigmoidal model (see Fig. 5.23a,b) is

$$y = \alpha + \beta \log[(X + \gamma)/(\gamma + \delta - X)] \qquad (5.4.25)$$

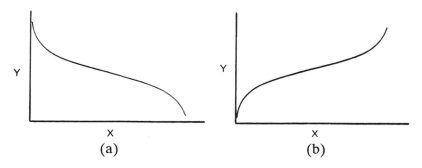

(a)   (b)

Figure 5.23a,b

Even when this model fits data rather well, so that there is only a relatively small amount of random variation of the data about the regression line, the estimation behavior of (5.4.25) is poor, with two or more of the parameters being far-from-linear. The use of this model is not recommended.

Another four-parameter model (see Fig. 5.24a,b) with a sigmoidal shape is

$$y = \alpha + \beta(1 - \gamma^X)^\delta \qquad (5.4.26)$$

This model tends to be unstable, with two or more of its parameters exhibiting considerable nonlinear behavior. Much better behavior results when δ can be fixed to have a constant value. There is virtually no difference in behavior between (5.4.26) and its parameterization

$$y = \alpha + \beta[1 - \exp(-\nu X)]^\delta \qquad (5.4.27)$$

The use of either of these four-parameter model functions is not recommended.

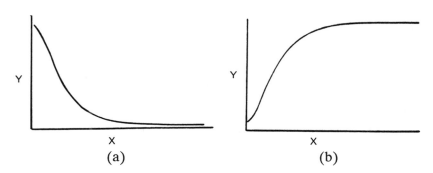

Figure 5.24

5.5 FIVE-PARAMETER CURVES

An extension of the four-parameter logistic (5.3.1) or of the Richards curve (5.4.12) is

$$y = \epsilon + \alpha/[1 + \exp(\beta - \gamma X)]^{1/\delta} \qquad (5.5.1)$$

or, in another parameterization where δ appears twice,

$$y = \epsilon + \alpha/\{1 + \delta \exp[-\beta(X - \gamma)]\}^{1/\delta} \qquad (5.5.2)$$

For reasons given in Section 5.4 in the discussion about (5.4.12), disastrous statistical properties in estimation are to be expected from either of these parameterizations.

Similarly, a model such as

$$y = \epsilon + \alpha \exp\{-[(X + \beta)/\gamma]^{\delta}\} \qquad (5.5.3)$$

tends to be very unstable, as a five-parameter model is unlikely to be a parsimonious model for fitting a sigmoidal curve. Its use is not recommended.

6

Models with One X Variable, Curves with Maxima and Minima

This chapter concerns models with a single explanatory variable in which the graph of Y versus X has at least one maximum or one minimum. In addition, the curve may have one or more inflection points.

6.1 ONE-PARAMETER CURVES

A model with a single parameter (see Figure 6.1) that has multiple maxima, minima, and inflection points is the trigonometric function

$$(*) \quad y = \cos(X + \alpha) \tag{6.1.1}$$

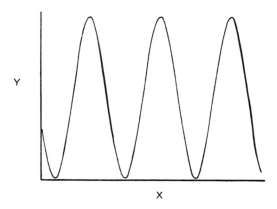

Y

X

Figure 6.1

with the parameter α measured in radians. There is very little nonlinear estimation behavior for (6.1.1). A similar model is

$$(*) \quad y = \sin(X + \beta) \qquad\qquad (6.1.2)$$

which has identical estimation properties to that of (6.1.1), with the least-squares estimate of β in (6.1.2) being shifted by $\pi/2(= 90°)$ from that of α in (6.1.1).

Another one-parameter model (see Fig. 6.2) with a maximum and an inflection point is

$$(*) \quad y = (1 + \alpha X)\exp(-\alpha X) \qquad\qquad (6.1.3)$$

the maximum occurring at $X = 0$ and the inflection point at $X = 1/\alpha$. A similar model (see Fig. 6.3), which enables the maximum to occur at a positive X value, is

$$(*) \quad y = \alpha X \exp(-\alpha X) \qquad\qquad (6.1.4)$$

Both (6.1.3) and (6.1.4) are close-to-linear in estimation behavior.

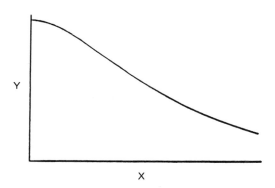

Figure 6.2

A further model with one parameter (see Fig. 6.4) is

$$(*) \quad y = 1 - \exp(-\alpha X^2) \tag{6.1.5}$$

which has a minimum at $X = 0$ and an inflection point at $X = (1/2\alpha)^{1/2}$.

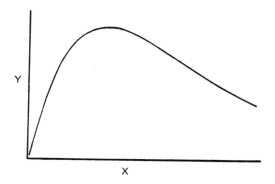

Figure 6.3

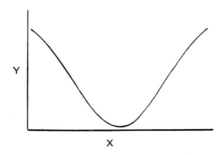

Figure 6.4

A similar model (see Figure 6.5), having a maximum rather than a minimum at $X = 0$, is

$$(*) \quad y = \exp(-\alpha X^2) \qquad\qquad (6.1.6)$$

There is very little nonlinearity in α in both (6.1.5) and (6.1.6).

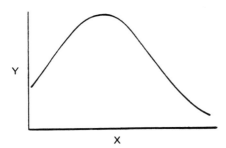

Figure 6.5

6.2 TWO-PARAMETER CURVES

A two-parameter model (see Figure 6.6) with multiple maxima, minima, and inflection points is the trigonometric function

$$(*) \quad y = \alpha \cos(X + \beta) \tag{6.2.1}$$

The intrinsic nonlinearity of this model is zero, as it is capable of being reparameterized to a linear model. This follows from the trigonometric identity

$$\cos(X + \beta) = \cos X \cos \beta - \sin X \sin \beta$$

In consequence, (6.2.1) becomes

$$(***) \quad y = \gamma \cos X + \delta \sin X \tag{6.2.2}$$

where $\gamma = \alpha \cos \beta$ and $\delta = -\alpha \sin \beta$. Since γ and δ are both linear-appearing parameters, (6.2.2) is a linear model.

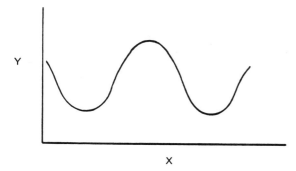

Figure 6.6

Another two-parameter model (see Figure 6.7) that has a single maximum and a single inflection point is

$$(*) \quad y = \alpha X^{\beta} \exp(-X) \qquad\qquad (6.2.3)$$

The behavior of (6.2.3) is reasonably close-to-linear. However, as β tends to be the worse-behaving parameter, it can be replaced by an expected-value parameter y_1, corresponding to $X = X_1$, to yield

$$(*) \quad y = \alpha \exp(-X)[y_1/\alpha \exp(-X_1)]^{\log X / \log X_1} \qquad\qquad (6.2.4)$$

Another two-parameter model (see Figure 6.8) with a maximum (which occurs at $X = 0$) and a pair of inflection points occurring at $X = \pm[1/(1+2\beta)]^{1/2}$ is

$$(*) \quad y = \alpha/(1+X^2)^{\beta} \qquad\qquad (6.2.5)$$

Although the behavior of (6.2.5) is reasonably close-to-linear, β, the worse-behaving of the two parameters, may be replaced by an

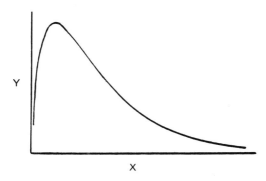

Figure 6.7

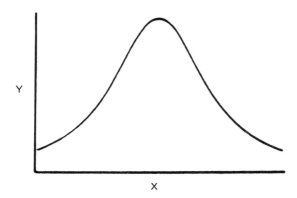

Figure 6.8

expected-value parameter y_1, corresponding to $X = X_1$, to yield

$$(*) \quad y = \alpha/(\alpha/y_1)^{\log(1+X^2)/\log(1+X_1^2)} \qquad (6.2.6)$$

A model with two parameters (see Fig. 6.9) having a minimum and a single inflection point is

$$(*) \quad y = (\beta - \alpha X)\exp(-\alpha X) \qquad (6.2.7)$$

The minimum occurs at $(\beta + 1)/\alpha$ and the inflection at $(\beta + 2)/\alpha$. The model is reasonably close-to-linear. To obtain a model (see Figure 6.10) with a maximum and a single inflection point, the parameters in the first expression in parentheses in (6.2.7) need to have their signs reversed, that is

$$(*) \quad y = (\alpha X - \beta)\exp(-\alpha X) \qquad (6.2.8)$$

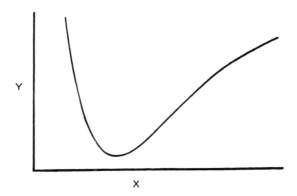

Figure 6.9

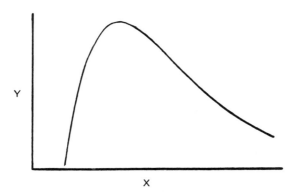

Figure 6.10

6.3 THREE-PARAMETER CURVES

A simple curve (see Figure 6.11 a,b) with a single maximum or minimum but without an inflection point is the quadratic polynomial

$$(***)\quad y = \alpha + \beta X + \gamma X^2 \tag{6.3.1}$$

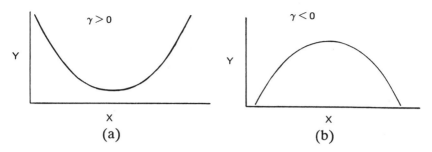

Figure 6.11

with the minimum or maximum occurring at $X = -\beta/2\gamma$. As (6.3.1) is a linear regression model, the estimators of α, β, and γ must be unbiased, normally distributed, minimum variance estimators (provided the error term satisfies the assumption of having an independent and identically distributed normal distribution).

A much more versatile three-parameter model (see Figure 6.12 a,b) than (6.3.1) is the inverse polynomial model

$$(*) \quad y = 1/(\alpha + \beta X + \gamma X^2) \tag{6.3.2}$$

which has already appeared as the Holliday (1960) model (4.3.18), in which the parameters α, β, and γ all exceed zero. Versatility in (6.3.2) is obtained by allowing some or all of its parameters to be

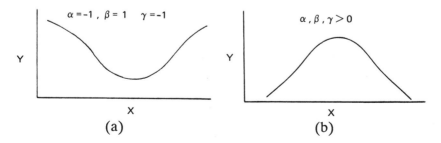

Figure 6.12

negative. A maximum or minimum occurs when $X = -\beta/2\gamma$, and two real inflection points are obtained when $4\alpha\gamma - \beta^2$ is greater than or equal to zero. As an example (see Figure 6.12a), if $\alpha = -1$, $\gamma = -1$, and $\beta = 1$, a minimum is obtained at $X = 0.5$, and inflection points appear at $X = 0$ and $X = 1$. A minimum or maximum may also occur when all three parameters are positive, as in Figure 6.12b.

Another three-parameter model (see Figure 6.13) is

$$y = (\alpha + \beta X)\gamma^X \qquad (6.3.3)$$

There is very little difference in statistical behavior between this form and its reparameterization

$$y = (\alpha + \beta X)\exp(-\delta X) \qquad (6.3.4)$$

with γ in (6.3.3) [or δ in (6.3.4)] being the worst-behaved parameter of the three parameters α, β, and γ, [or α, β, and δ]. The term γ [or δ] can be replaced by an expected-value parameter y_1, corresponding to $X = X_1$, to give

$$(*) \quad y = (\alpha + \beta X)[y_1/(\alpha + \beta X_1)]^{X/X_1} \qquad (6.3.5)$$

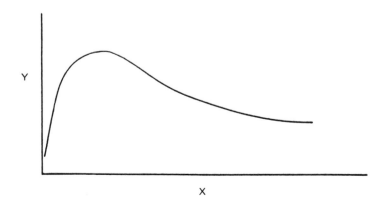

Figure 6.13

This model has good statistical properties.

   Another three-parameter model (see Figure 6.14), with the capability of having a maximum and an inflection point, is

$$(*) \quad y = X/(\alpha + \beta X + \gamma X^2) \tag{6.3.6}$$

This is really the same model as (5.3.16) of Chapter 5, except that the sign of parameter γ has been changed. This model tends to have good estimation properties.

   A further three-parameter curve (see Fig. 6.15) is

$$(*) \quad y = \alpha + \beta \cos(X + \gamma) \tag{6.3.7}$$

By use of the trigonometric identity given in Section 6.2, it is easily shown that (6.3.7) may be reparameterized to the linear model

$$(***) \quad y = \alpha + \delta \cos X + \epsilon \sin X$$

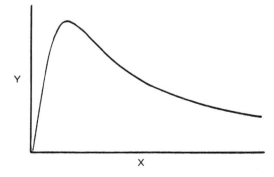

Figure 6.14

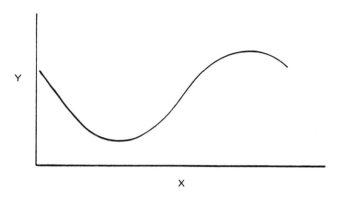

Figure 6.15

Another three-parameter model (see Figure 6.16 a,b), which was already considered in Section 4.3 as (4.3.39), is

$$(*) \quad y = \exp(\alpha + \beta X + \gamma X^2) \tag{6.3.8}$$

This model can have a maximum or a minimum at $X = -\beta/2\gamma$, depending on the sign of γ, and a pair of inflection points when there is a maximum. All parameters, α, β, and γ, in (6.38) have similar behavior in estimation, and all are close-to-linear. Hence,

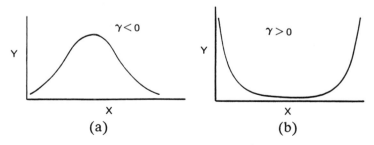

(a)   (b)

Figure 6.16

(6.3.8) is a very versatile model, capable of fitting curves with a maximum or a minimum as well as convex/concave curves.

Another three-parameter model (see Figure 6.17), which was studied in Section 4.3 as (4.3.15), is the extended Freundlich model (Sibbesen, 1981),

$$(*) \quad y = \alpha X^{\beta X^{-\gamma}} \tag{6.3.9}$$

It is capable of having a maximum or minimum at $X = \exp(1/\gamma)$ and a pair of inflection points. If α or β is negative, a minimum is obtained. The statistical properties in estimation of all parameters of (6.3.9) are very good when the data contain sufficiently large X values so that α can be determined accurately. Otherwise, the behavior of the estimator of α is not close-to-linear, and α is best replaced by an expected-value parameter y_1, corresponding to $X = X_1$, to give

$$(*) \quad y = y_1 X^{\beta X^{-\gamma}} / X_1^{\beta X_1^{-\gamma}} \tag{6.3.10}$$

which has already appeared in Section 4.3 as (4.3.16).

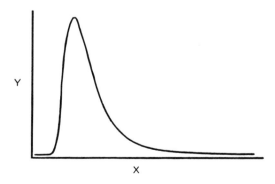

Figure 6.17

Yet another three-parameter model (see Figure 6.18), which was considered in Section 4.3 as (4.3.17), is the Gunary (1970) model,

$$(*) \quad y = X/(\alpha + \beta X + \gamma\sqrt{X}) \qquad\qquad (6.3.11)$$

This model can have a maximum or minimum at $X = 4\alpha^2/\gamma^2$ and a pair of inflection points. The three parameters α, β, and γ in (6.3.11) are all reasonably close-to-linear in estimation behavior.

A model (see Figure 6.19) that is similar to (6.3.11) in that it can have a maximum or a minimum and a pair of inflection points is

$$(*) \quad y = \alpha X^\beta \exp(-\gamma X) \qquad\qquad (6.3.12)$$

The statistical properties of (6.3.12) in estimation are close-to-linear. If the error term for (6.3.12) is multiplicative rather than additive, which is equivalent to taking the logarithm of the right-hand side of (6.3.12) and replacing the left-hand side with the expectation of the logarithm of Y, it is clear that the resulting

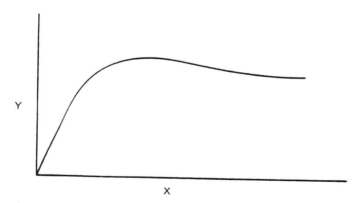

Figure 6.18

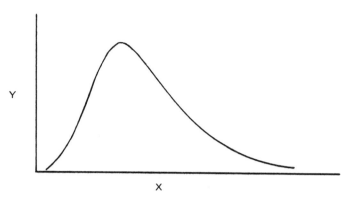

Figure 6.19

expression,

$$E(\log Y) = \log\alpha + \beta\log X - \gamma X$$

can be made into a linear model by replacing $\log\alpha$ with another parameter δ (say).

An alternative parameterization to (6.3.12) with similar estimation properties is

$$(*) \quad y = X^\beta \exp(\alpha - \gamma X) \qquad\qquad (6.3.13)$$

Another three-parameter model (see Figure 6.20) with a maximum (at $X = 0$) and a pair of inflection points is

$$y = \alpha/(1 + \beta X^2)^\gamma \qquad\qquad (6.3.14)$$

This model tends to have very poor estimation behavior, with β and γ both being far-from-linear. Its use is not recommended.

A further model (see Figure 6.21 a,b) with three parameters is one that has been suggested by the probability distribution known

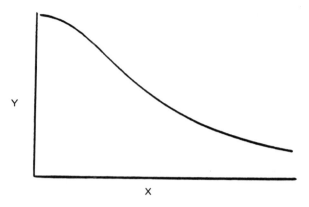

Figure 6.20

as the beta distribution,

$$y = \alpha X^{\beta}(1-X)^{\gamma} \tag{6.3.15}$$

The range of the explanatory variable X is 0 to 1, the value of $\beta/(\beta + \gamma)$ determining the position of the maximum. This model does not have particularly good estimation properties and must be used with caution.

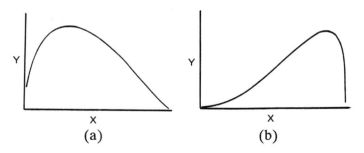

(a)   (b)

Figure 6.21

6.4   FOUR-PARAMETER CURVES

A four-parameter model (see Figure 6.22 a,b) having a single maximum or minimum, a single inflection point, and an asymptote is

$$(*)\quad y = (\alpha + \beta X)\gamma^X + \delta \tag{6.4.1}$$

A maximum (or minimum) occurs at $X = -1/\log\gamma - \alpha/\beta$ and an inflection point at $X = -2/\log\gamma - \alpha/\beta$.

A reparameterization of (6.4.1)

$$y = (\alpha + \beta X)\exp(-\epsilon X) + \delta \tag{6.4.2}$$

is generally not as good as (6.4.1). The worst-behaving parameter in (6.4.1) [and (6.4.2)] is often β, which may be replaced by an expected-value parameter y_1, corresponding to $X = X_1$, to yield

$$(*)\quad y = [\alpha(1 - X/X_1) + (X/X_1)(y_1 - \delta)/\gamma^{X_1}]\gamma^X + \delta \tag{6.4.3}$$

which has very good estimation properties.

Another four-parameter model (see Figure 6.23) is the Bragg equation (Bragg and Packer, 1962), which is based on the formula

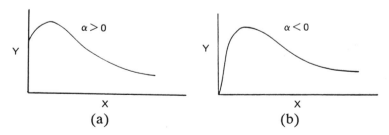

$$\alpha > 0 \qquad\qquad \alpha < 0$$

(a)   (b)

Figure 6.22

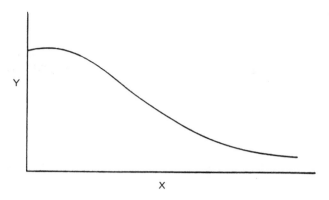

Figure 6.23

for the normal (Gaussian) distribution

$$(*) \quad y = \alpha + \beta \exp[-\gamma(X - \delta)^2] \tag{6.4.4}$$

This model is capable of having a maximum, a minimum, and inflection points.   Model (6.4.4) was found to fit data on the preferred orientation of carbon materials very well (Horton, 1979) and had excellent statistical properties (Ratkowsky, 1986d).   .

Another model (see Figure 6.24) with similar capabilities to (6.4.4) is what Horton (1979) referred to as the Lorentz equation,

$$y = \alpha + \beta/[1 + \gamma(X - \delta)^2] \tag{6.4.5}$$

Model (6.4.5) had poorer statistical properties than (6.4.4) for carbon materials and didn't fit data as well as (6.4.4) or two models to follow, (6.4.6) and (6.4.7).

Another model (see Figure 6.25) capable of fitting similar data to (6.4.4) and (6.4.5) is the Bacon (1956) equation

$$y = \alpha + \beta[\cos(X - \delta)]^\gamma \tag{6.4.6}$$

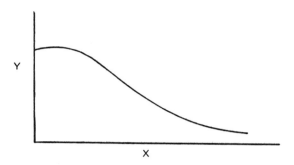

Figure 6.24

containing the trigonometric function cosine. Model (6.4.6) generally exhibited good statistical properties, with the exception of one data set on carbon materials (Ratkowsky, 1986d).

Yet another model (see Figure 6.26) with fitting abilities similar to (6.4.4–6.4.6) is the Ergun (1976) equation

$$y = \alpha + \beta \exp[-\gamma \sin^2 (X - \delta)] \tag{6.4.7}$$

which contains the trigonometric function sine. For carbon materials, (6.4.7) was almost as good as equation (6.4.4) insofar as

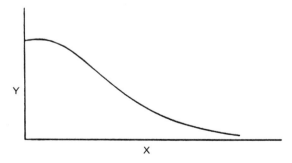

Figure 6.25

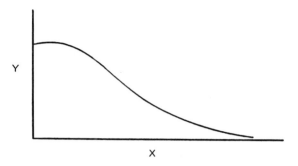

Figure 6.26

goodness-of-fit and statistical behavior were concerned, except for
its performance on one data set (Ratkowsky, 1986d).

One further model (see Figure 6.27) with a fitting capability
similar to (6.4.4–6.4.7) is the Ruland (1967) equation

$$y = \alpha + \beta/[1 + \gamma\sin^2(X - \delta)] \tag{6.4.8}$$

which, like (6.4.7), incorporates the trigonometric function sine.
For carbon materials, it was found (Ratkowsky, 1986d) that this
model did not fit data on carbon materials as well as (6.4.4), (6.4.6),
and (6.4.7).

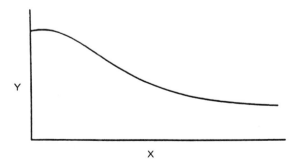

Figure 6.27

Another four-parameter curve (see Figure 6.28) with maxima, minima, and inflection points is

$$(*) \quad y = \alpha \cos(X + \gamma) + \beta \cos(2X + \gamma) + \delta$$

or

$$y = \alpha \sin(X + \gamma) + \beta \sin(2X + \gamma) + \delta \qquad (6.4.9)$$

The amount of nonlinear behavior exhibited by the estimators of the parameters of (6.4.9) is very small. Hence, (6.4.9) is close-to-linear.

Another model (see Figure 6.29) with four parameters is

$$y = \alpha - \beta \gamma^X - \delta X \qquad (6.4.10)$$

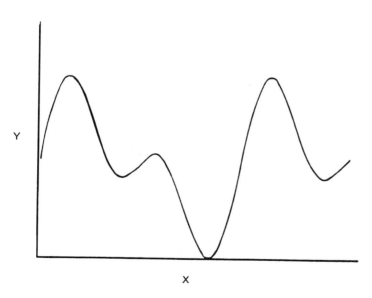

Figure 6.28

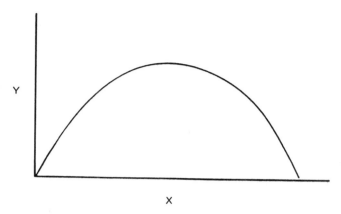

Figure 6.29

which is made up of the asymptotic regression model (4.3.1) minus a linear term. The model is capable of having a maximum but no inflection points. This feature limits the versatility of the model, and (6.4.10) also exhibits poor behavior in estimation, as all parameters are far-from-linear in behavior.

Another four-parameter model (see Figure 6.30) that is capable of having a maximum, a minimum, and inflection points is the

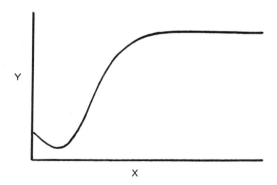

Figure 6.30

rational functions model

$$(*) \quad y = (\beta + \gamma X)/(1 + \alpha X + \delta X^2) \tag{6.4.11}$$

which is an extension of (4.3.24). The behavior of (6.4.11) in estimation is reasonably good.

Another four-parameter model (see Figure 6.31) with maxima, minima, and inflection points is

$$(*) \quad y = \alpha \cos(X + \delta) + \beta \cos(2X + \delta) + \gamma \cos(3X + \delta) \tag{6.4.12}$$

The nonlinearity in (6.4.12) is due to the presence of the parameter δ in this example of a truncated Fourier series, but the extent of nonlinear behavior in estimation is very small.

Another four-parameter model (see Figure 6.32 a,b) with the capability of having a maximum (or a minimum) and an inflection

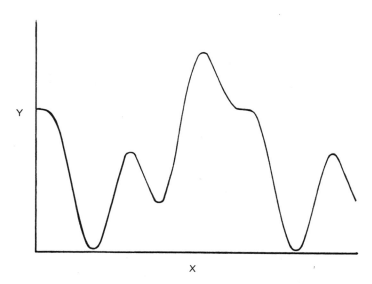

Figure 6.31

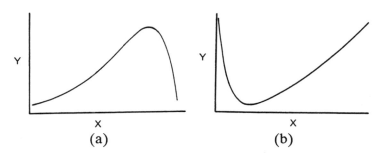

Figure 6.32

point is

$$y = \alpha \exp(\beta X) + \gamma \exp(\delta X) \tag{6.4.13}$$

which is, in fact, the same model as (4.4.4), except for some changes in the signs of the parameters. Generally, a minimum or a maximum is attainable when an odd number (that is, 1 or 3) of the signs of α, β, γ, and δ are negative. The magnitudes of the parameters then determine whether the stationary point of the curve is a maximum or a minimum. As noted in the discussion of (4.4.4), the statistical behavior of the parameterization given by (4.4.4), and hence also (6.4.13), is often far-from-linear, and one may have to resort to using a parameterization with two expected-value parameters y_1 and y_2, corresponding to $X = X_1$, and $X = X_2$, respectively, that is,

$$y = \frac{y_1(\lambda_1^X \lambda_2^{X_2} - \lambda_2^X \lambda_1^{X_2}) - y_2(\lambda_1^X \lambda_2^{X_1} - \lambda_2^X \lambda_1^{X_1})}{\lambda_1^{X_1} \lambda_2^{X_2} - \lambda_1^{X_2} \lambda_2^{X_1}} \tag{6.4.14}$$

in which $\lambda_1 = \exp(\beta)$ and $\lambda_2 = \exp(\delta)$. The estimation behavior of y_1 and y_2 is excellent, but that of λ_1 and λ_2 may not be. Hence, (6.4.14) needs to be used with caution.

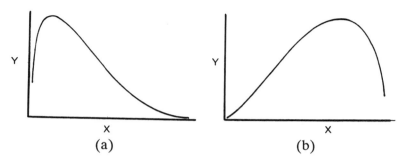

Figure 6.33

A further four-parameter model (see Figure 6.33a,b) with a maximum and a pair of inflection points is

$$y = \alpha X^\beta (\delta - X)^\gamma \qquad (6.4.15)$$

This model can be viewed as an extension of (6.3.15), with X now having the capability of ranging between 0 and δ instead of being confined to the range 0–1. The maximum occurs at $X = \beta\delta/(\beta+\gamma)$. However, this model tends to have very poor estimation behavior and must be used with caution, as several parameters may simultaneously be far-from-linear.

The models considered in this section do not, by any means, exhaust the almost unlimited possibilities of four-parameter functional forms that have maxima, minima, and inflection points. The more complicated models, however, tend to exhibit poor statistical estimation properties. Some of these possibilities are considered in Section 8.1.

6.5   FIVE-PARAMETER CURVES

The five-parameter rational functions model (see Fig. 6.34),

$$(*) \quad y = (\beta + \gamma X + \epsilon X^2)/(1 + \alpha X + \delta X^2) \qquad (6.5.1)$$

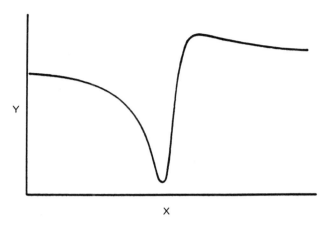

Figure 6.34

which is an extension of (4.3.24) and (6.4.11), is capable of having a maximum and a minimum, as well as inflection points.
The statistical properties of (6.5.1) can be very good and the
model very versatile, mimicking many other mathematical functions (Ratkowsky, 1987). There are other circumstances, however,
in which (6.5.1) can exhibit far-from-linear behavior, even though
the model may be fitting the data rather well. This is likely to be
the result of overparameterization, that is, using a model with too
many parameters (five in this case) when one with fewer parameters might have done as well, or almost as well (see Sec. 8.2).
Extremely high parameter correlation is an almost certain diagnostic indicative of over-parameterization (see Sec. 2.5.1).

When (6.5.1) is a suitable model for one's data, with good
estimation properties, the same note of caution about the use of
(6.5.1) applies here as applied to the use of (4.3.24). It is possible
to have singularities in the denominator of (6.5.1), occurring when
$X = [-\alpha \pm (\alpha^2 - 4\delta)^{1/2}]/2\delta$, in which case a vertical asymptote
appears at these values of X.

7

Models with More than One
Explanatory Variable

7.1 INTRODUCTION

Models with more than one explanatory variable are much more
difficult to classify than models with a single explanatory variable,
which could be classified as convex/concave (Chap. 4), sigmoidal
(Chap. 5), or having maxima or minima (Chap. 6). In addition,
models with two or more explanatory variables are generally more
specialized with respect to their applications than those with but
a single explanatory variable and are therefore less likely to be
employed in more than one discipline. Therefore, this chapter
only scratches the surface of the myriad of possible models with
more than one explanatory variable. Because of the "curse of
dimensionality" discussed in Section 2.5.4, models with more
than four or five parameters are seldom likely to be "close-to-
linear" nonlinear regression models, as discussed in Section 2.1.

However, models with more than one explanatory variable need not necessarily have many parameters. The examples in this chapter contain four parameters or less.

7.2 THE NERNST EQUATION

The Nernst equation describes the relationship between the potential of a specific ion electrode and a chemical substance that is present naturally in a system as well as being added by the experimenter to the system by means of a standard addition procedure. For example, suppose one wishes to determine the extent to which a certain compound is present in stored fish. If an extract in solution form is taken from a fish, varying amounts X_1 of the compound are added, and potentiometric measurements Y are made with a specific ion electrode. If the final volume of the solution after each addition is X_2, the potential Y is related to X_1 and X_2 by the nonlinear regression model (see Fig. 7.1a,b)

$$(*) \quad y = \alpha - \beta[\log(X_1 + \gamma) - \log X_2] \tag{7.2.1}$$

Model (7.2.1) is one of the forms of the Nernst equation and has been used in the potentiometric estimation of ammonia in blood (Meyerhoff and Robins, 1980). Figures 7.1a and 7.1b plot y against X_1 only, and y against X_2 only, respectively, because of the

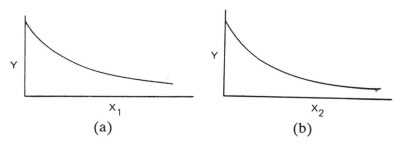

(a)   (b)

Figure 7.1

difficulty of depicting the response plus explanatory variables in three dimensions. For 22 sets of data on trimethylamine in various species of fish (data courtesy Mr. P. Kearney), only one data set had far-from-linear behavior in combination with (7.2.1). For the other 21 data sets, (7.2.1) exhibited close-to-linear estimation behavior.

7.3 CATALYTIC CHEMICAL AND BIOCHEMICAL REACTIONS

Section 2.3.2 introduced the principle of putting the parameters in the denominator of the expression relating the reaction rate of a chemical or biochemical reaction and the explanatory variables. In (2.3.16), the parameters k, K_A, K_B, the first being a rate constant and the second and third being equilibrium constants, appear in the numerator of the expression as the product of the three constants (K_A and K_B also appear in the denominator of the expression, but that is incidental). By dividing numerator and denominator by kK_AK_B, new parameters ϕ_1, ϕ_2, and ϕ_3 are obtained that appear only in the denominator of the expression [see (2.3.17)]. This principle was applied by Ratkowsky (1985) to models of the Hougen and Watson (1947) formalism. Examples of such expressions having all their parameters in the denominator are (see Ratkowsky, 1985, for details of the sources of these examples)

$$r = X_1X_2/(\phi_1 + \phi_2X_1 + \phi_3X_2 + \phi_4X_3)^2 \tag{7.3.1}$$

$$r = (X_1^2 - X_2^2/K)/(\phi_1 + \phi_2X_1 + \phi_3X_2)^2 \tag{7.3.2}$$

$$r = (X_1^2 - X_2X_3/K)/(\phi_1 + \phi_2X_1 + \phi_3X_2 + \phi_4X_3)^2 \tag{7.3.3}$$

$$r = X_1X_2/(\phi_1X_1 + \phi_2X_2X_3) \tag{7.3.4}$$

$$r = X_1/(\phi_1 + \phi_2X_1 + \phi_3X_2) \tag{7.3.5}$$

and

$$r = (X_2 - X_3/K)/(\phi_1 + \phi_2X_1 + \phi_3X_2 + \phi_4X_4) \tag{7.3.6}$$

In (7.3.1–7.3.6), r is the response variable, X_1, X_2, X_3, and X_4 are explanatory variables, and K, when present, represents a known equilibrium constant. With few exceptions, all parameters in these models (ϕ_1, ϕ_2, ϕ_3, and ϕ_4) exhibited close-to-linear estimation behavior (Ratkowsky, 1985). It should be obvious that the number of explanatory variables, which range from two to four in (7.3.1–7.3.6), is of no consequence in itself in influencing the behavior of the parameters.

The same principle of putting the parameters in the denominator applies to biochemical reactions as well as to chemical reactions. In Section 4.2, it was asserted that the Michaelis–Menten model of enzyme kinetics (4.2.17),

$$y = y_{max}X/(X + X_K)$$

could be reparameterized by dividing numerator and denominator by y_{max} to give (4.2.18),

$$(*) \quad y = X/(\theta_1 X + \theta_2)$$

For more complicated enzyme reaction equations, for example,

$$y = y_{max}K_aK_bC_1C_2/(1 + K_aC_1 + K_aK_bC_1C_2 + K_aK_cC_1C_3)$$
$$(7.3.7)$$

where the explanatory variable C_1, C_2, and C_3 are the concentrations of the substances involved in the reaction and y_{max}, K_a, K_b, and K_c are the parameters, one divides through by the product $y_{max}K_aK_b$ to give

$$(*) \quad y = C_1C_2/(\theta_1 + \theta_2C_1 + \theta_3C_1C_2 + \theta_4C_1C_3) \qquad (7.3.8)$$

where the "new" parameters θ_1, θ_2, θ_3, and θ_4 exhibit close-to-linear behavior.

7.4   THE VON BERTALANFFY CAPTURE–
RECAPTURE MODEL

It was noted in Section 4.3 that the asymptotic regression model
(4.3.1–4.3.9) was also known as the von Bertalanffy law in fisheries
research. In that discipline, the model relates the length of a fish
(or school of fish) to its age. Another form of data gathering
involves capturing individuals of a species of fish or shellfish,
measuring its initial length, tagging and releasing the individuals
to be recaptured later, when the length of each individual is
remeasured. The age of each individual is not known, but it is
easy to derive a two-parameter expression relating the increment
in length to the increment in time between initial capture and
recapture (for example, see Fabens, 1965, for the derivation). The
expression is

$$\Delta y = (\alpha - l_1)[1 - \exp(-k\Delta T)] \qquad\qquad (7.4.1)$$

where Δy is the increment in length during the time at liberty
ΔT, l_1 the length at initial capture, and α the asymptotic length
characteristic of the school of fish from which the individuals are
drawn (see Fig. 4.22). There are two explanatory variables in this
model, l_1 and ΔT.

The properties in estimation of (7.4.1) are generally poor, with
considerable bias being experienced by the estimators of k and
α. However, as shown by Francis (1988a), α and k may each
be replaced by expected-value parameters. Since there are two
explanatory variables, the user must choose arbitrary values for
each variable for each expected-value parameter. For one of
these variables, the same arbitrary value may be chosen for both
expected-value parameters. Although the choice of the values of
the explanatory variables is arbitrary, the values should be chosen
to lie within the observed ranges of each of these variables in order
that the estimators of the expected-value parameters exhibit close-
to-linear behavior. For l_1, a suitable choice might be the smallest
and largest initial lengths. For ΔT, one can choose a single value

that falls close to the mean or median of the observed time between release and recapture. Thus, if l_{11} is one chosen value of l_1, with ΔT_m being the chosen value of the time increment, one obtains, from (7.4.1),

$$\Delta y_1 = (\alpha - l_{11})[1 - \exp(-k\Delta T_m)]$$

Similarly, if l_{12} is the other chosen value of l_1, with ΔT_m once again being the chosen value of the time increment, one obtains

$$\Delta y_2 = (\alpha - l_{12})[1 - \exp(-k\Delta T_m)]$$

From these two equations, the original parameters α and k are easily solved for, in terms of the expected-value parameters Δy_1 and Δy_2, to yield, after substitution into (7.4.1),

$$(*) \quad \Delta y = \left[\frac{l_{12}(\Delta y_1) - l_{11}(\Delta y_2)}{\Delta y_1 - \Delta y_2} - l_1\right]\left[1 - \left(1 - \frac{\Delta y_1 - \Delta y_2}{l_{12} - l_{11}}\right)^{\Delta T/\Delta T_m}\right]$$

$$(7.4.2)$$

The parameters Δy_1 and Δy_2 are mean growth rates for fish of arbitrary lengths l_{11} and l_{12}, respectively. If the reference time increment ΔT_m is one year, for example, Δy_1 and Δy_2 are mean annual growth rates. If ΔT_m is one month, then the growth rates are mean monthly growth rates, and so forth.

This section has illustrated the fact that the principles of expected-value parameterizations described in Section 2.3.1 need not be confined to models with only a single explanatory variable.

8

Other Models and Excluded Models

Chapters 4–7 dealt with a wide variety of nonlinear regression models, but the models are only a fraction of the nonlinear models that are possible. It is hoped that all the commonly used models have been included in those chapters. In this chapter, we will examine a number of additional nonlinear regression models, some of which were excluded from Chapters 4–7 because it was clear that those models were not likely to find a place in the modeler's stockpile of potentially useful models. Section 8.1 deals with models that result from the solutions of 32 differential equations advanced by Watt (1968). Section 8.2 deals with the harmful effects of overparameterization, and Section 8.3 is devoted to a discussion of a few other models that have been excluded from Chapters 4–7.

8.1 MODELS FROM DIFFERENTIAL EQUATIONS

A sound basis for finding models that may be useful in resource management was advocated by Watt (1968). He proposed 32 differential equations, in which the derivative dY/dX of the response Y with respect to a single explanatory variable X was the product of five elements, some or all of which may be present. The five elements were (1) a linear term in X, that is, $b + cX$, where b and c are constants, (2) Y, (3) $1/X$, (4) $Y_{max} - Y$, where Y_{max} is the maximum possible response, and (5) $1/(X - X_{min})$, where X_{min} is the minimum possible value of X. The differential equation containing all five elements is

$$\frac{dY}{dX} = \frac{(b + cX)Y(Y_{max} - Y)}{X(X - X_{min})} \tag{8.1.1}$$

which has as its solution

$$y = Y_{max}/\{1 + \alpha X^{\beta}/[(X - X_{min})^{\gamma}]\} \tag{8.1.2}$$

the equation containing five parameters, α, β, γ, Y_{max}, and X_{min}.

The statistical properties of (8.1.2) tend to be poor, as they are for other models for which the product of elements $X(X - X_{min})$ appear in the denominator of the differential equation (8.1.1). Indeed, if X cannot be less than a certain value X_{min}, it doesn't seem sensible that X should appear in the denominator of (8.1.1) as a multiplier of the element $X - X_{min}$. In other words, $(X - X_{min})^2$ might be a more sensible term to have in the denominator of the differential equation than the product $X(X - X_{min})$. Of the 32 models formed by combination of the elements considered above, eight contained the combination $X(X - X_{min})$ and are excluded from further consideration in this handbook on two grounds: (1) they may not be sensible models to consider; (2) their statistical properties in estimation tend to be poor. The models thus excluded, in addition to (8.1.2), are

$$y = \alpha/[X^\beta(X - X_{min})^\gamma] \tag{8.1.3}$$

$$y = Y_{max} - \alpha X^\beta/(X - X_{min})^\gamma \tag{8.1.4}$$

$$y = \log[\alpha(X - X_{min})^\gamma/X^\beta] \tag{8.1.5}$$

$$y = Y_{max}/\{1 + \alpha[X/(X - X_{min})]^\beta\} \tag{8.1.6}$$

$$y = \alpha[X/(X - X_{min})]^\beta \tag{8.1.7}$$

$$y = Y_{max} - \alpha[X/(X - X_{min})]^\beta \tag{8.1.8}$$

and

$$y = \alpha + \beta\log[(X - X_{min})/X] \tag{8.1.9}$$

Several other models among the 32 models obtained by Watt (1968) did not exhibit good estimation properties. These included

$$y = \delta/\{1 + \exp[-(\alpha + \beta X + \gamma X^2)]\} \tag{8.1.10}$$

$$y = \exp(\alpha + \beta X)/(X - \gamma)^\delta \tag{8.1.11}$$

$$y = \delta - \exp(\alpha + \beta X)/X^\gamma \tag{8.1.12}$$

$$y = \delta - \exp[-(\alpha + \beta X + \gamma X^2)] \tag{8.1.13}$$

and

$$y = \delta/[1 + \exp(\alpha + \beta X)/X^\gamma] \tag{8.1.14}$$

One of the lessons to be learned from this is that a model should not be "overloaded." For example, (8.1.11), (8.1.12), and (8.1.14) combine an exponential term containing a linear expression in X with another term containing a power (exponent) of X. This combination seems to lead to poor estimation behavior and should be avoided in general. The one exception was (6.3.12) of Section 6.3,

$$y = \alpha X^\beta \exp(-\gamma X)$$

which exhibited close-to-linear estimation behavior. That model, however, is much simpler than (8.1.11), (8.1.12), and (8.1.14).

Model (8.1.10) is an extension of the well-behaved three-parameter logistic model (5.3.1) by the addition of the term γX^2. Once again, the poor statistical behavior of (8.1.10) probably stems from trying to overload the model by combining two elements, namely, (1) a logistic expression and (2) a polynomial expression. Similarly, the relatively poor statistical behavior of (8.1.13), compared with (6.3.8) of Section 6.3,

$$y = \exp(\alpha + \beta X + \gamma X^2)$$

which exhibited close-to-linear estimation behavior, is probably due to the inclusion of the extra parameter δ, as it is that parameter that is the worst-behaving parameter of the set of four parameters.

Another model that resulted from the differential equations of Watt (1968) is

$$y = \alpha - \beta/(X - \gamma)^\delta \tag{8.1.15}$$

This equation is similar to the generalized hyperbola (4.4.7) considered in Section 4.4. Its basic shape is that of a rectangular hyperbola with the origin of the X axis shifted so that there is an asymptote at $X = \gamma$ (see Fig. 8.1). The statistical properties in estimation of (8.1.15) are not particularly good, and close-to-linear estimation behavior results only when the model happens to fit the data very well. As was the case with (4.4.7), models such as (8.1.15) seem overparameterized for the task of fitting a rectangular hyperbola, whose basic model has but two parameters [see (4.2.16–4.2.19)]. In such cases, good estimation behavior cannot be expected unless the model fits the data extremely well, indicating that that model is tailor-made for that data set.

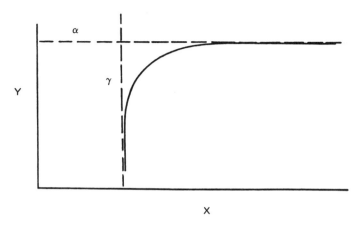

Figure 8.1

Another model that resulted from one of the differential equations of Watt (1968) is (see Fig. 8.2a,b)

$$y = \alpha + \beta X + \gamma \log(X + \delta) \tag{8.1.16}$$

This model can have a maximum or a minimum but no inflection point. In that respect, it is similar to (6.4.10) and, like that equation, does not have good statistical properties in estimation, with

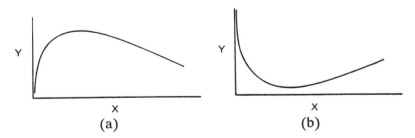

(a)   (b)

Figure 8.2

α and δ both being parameters with poor estimation properties. Although α or δ can individually be replaced by an expected-value parameter, there is no algebraic solution for α and δ in terms of the remaining parameters when one tries to replace both of them simultaneously.

In this section, we have excluded 16 of the 32 models contained in the differential equations of Watt (1968). The remaining 16 models contain some of the most useful linear and nonlinear regression models known to science, for example, the logistic model (see Sec. 5.4) and the asymptotic regression model (see Sec. 4.3), provided, of course, that parameterizations with good statistical properties are used. These models are scattered throughout Chapters 4–6.

8.2 THE CONSEQUENCES OF OVERPARAMETERIZATION

There are no hard-and-fast rules governing the question of how one might predict whether a nonlinear regression model is going to exhibit good estimation properties. However, the material in Section 8.1 does suggest that if a model is complicated in appearance (that is, combining more than one mathematical expression, such as an extended exponential term in the explanatory variable X with a power term in X) or not "sensible" (that is, containing both X and $X - X_{min}$, where X_{min} is a minimum value for X), it is likely to have poor estimation properties. Simplicity is always a desideratum, as emphasized in Section 2.5.4, and models that are likely to exhibit good estimation behavior are those that generally have (1) a simple form, and (2) few parameters.

One of the most serious causes of poor estimation behavior is overparameterization, that is, using a formula with more parameters than are needed. To understand this, it is necessary to consider the three-, four-, and five-parameter rational functions models considered in Chapters 4 and 6:

$$y = (\beta + \gamma X)/(1 + \alpha X) \tag{4.3.24}$$

$$y = (\beta + \gamma X)/(1 + \alpha X + \delta X^2) \tag{6.4.11}$$

and

$$y = (\beta + \gamma X + \epsilon X^2)/(1 + \alpha X + \delta X^2) \tag{6.5.1}$$

The three-parameter model (4.3.24) is capable only of fitting convex/concave data, whereas both (6.4.11) and (6.5.1) can fit data having a maximum and a minimum. Of course, the latter two equations can also fit convex/concave data, since the data may fit only a portion of the full curve, as in Figure 8.3. A concave portion is included between single vertical bars, and a portion having an inflection point is included between double vertical bars. Thus, models such as (6.4.11) and (6.5.1) are capable of fitting certain data sets whose range covers less than the full capabilities of those models. However, when models such as (6.4.11) and (6.5.1) are used for convex/concave data, poor statistical properties are likely to result. Two examples of how adding unnecessary parameters to a model can destroy the simpler model's good statistical properties without improving the goodness-of-fit substantially are given in Table 3 of Ratkowsky (1987). It is possible to devise some general rules for the use of a series of selected models such as (4.3.24),

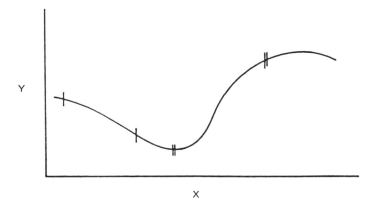

Figure 8.3

(6.4.11), and (6.5.1), where each successive model differs from the preceding model by the addition of an extra term.

Rule 1

Start the fitting process with the simplest model of the set that conforms to the characteristics of the data, that is, (4.3.24) for convex/concave data and (6.4.11) for data with inflection points, a maximum and/or a minimum.

Rule 2

Decide whether the goodness-of-fit is satisfactory by examining whether there are runs of like-signed residuals and employing other diagnostic tests mentioned in Section 1.4.

Rule 3

If the goodness-of-fit is not satisfactory, fit the next-most-complicated model, that is, (6.4.11) for convex/concave data and (6.5.1) for data with inflection points, a maximum, or a minimum. If this does not improve the goodness-of-fit, then the set of models under consideration is probably the wrong set, and different models should be considered.

Rule 4

If the goodness-of-fit is significantly improved, one should expect the estimation properties of the new model to remain good or to improve (see the last example in Ratkowsky, 1987, Table 3). Failure of the more complicated model to improve the goodness-of-fit will very probably result in a deterioration of the estimation behavior of the model.

A further model that has been excluded from consideration in Chapter 6 of this handbook is that of Sharpe and De Michele (1977) for the dependence of the rate of poikilotherm development on temperature, under nonlimiting substrate conditions, namely (see Fig. 8.4),

$$r = \frac{T \exp[(\phi - \Delta H_A^{\neq}/T)/R]}{1 + \exp[(\Delta S_L - \Delta H_L/T)/R] + \exp[\Delta S_H - \Delta H_H/T)/R]}$$

$$(8.2.1)$$

where r = rate of development, T = absolute temperature, R = universal gas constant, and the six parameters ϕ, $\Delta H_A^{\neq}$, ΔS_L, ΔH_L, ΔS_H, and ΔH_H are thermodynamic constants reflecting the organism's control enzyme system. The estimation behavior of (8.2.1) is very poor, as was recognized by Schoolfield et al. (1981). They attributed the poor estimation behavior to high correlation of the estimators and proposed the following reparameterization:

$$r = \frac{\rho\left(\frac{T}{298}\right)\exp\left[\left(\frac{\Delta H_A^{\neq}}{R}\right)\left(\frac{1}{298} - \frac{1}{T}\right)\right]}{1 + \exp\left[\left(\frac{\Delta H_L}{R}\right)\left(\frac{1}{T_{1/2_L}} - \frac{1}{T}\right)\right] + \exp\left[\left(\frac{\Delta H_H}{R}\right)\left(\frac{1}{T_{1/2_H}} - \frac{1}{T}\right)\right]}$$

$$(8.2.2)$$

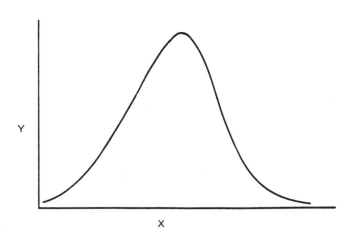

Y

X

Figure 8.4

where ρ, $\Delta H_A^{\neq}$, ΔH_L, $T_{1/2_L}$, ΔH_H, and $T_{1/2_H}$ are the six parameters to be estimated. Schoolfield et al. (1981) were able to fit several sets of data with (8.2.2), which they were unable to fit with (8.2.1).

Lowry and Ratkowsky (1983) pointed out that both (8.2.1) and (8.2.2) have an undesirably high degree of intrinsic and parameter-effects nonlinearity (see Sec. 2.2), leading to biased, nonnormally distributed parameter estimators. They showed that some difficulty could still be experienced in fitting the reparameterized model (8.2.2) and that its statistical properties were only a slight improvement over those of (8.2.1). Furthermore, they noted that the basic model was biologically incomplete as it failed to take account of irreversible enzyme reactions at both high and low temperatures, resulting in the prediction of a finite growth rate at all temperatures between $-\infty$ and $+\infty$. For the dependence of bacterial culture growth on temperature, Ratkowsky et al. (1983) presented a simpler four-parameter model (see Fig. 8.5),

$$(*) \quad \sqrt{r} = \beta(T - T_{\min})\{1 - \exp[\gamma(T - T_{\max})]\} \tag{8.2.3}$$

where $T_{\min}$ and $T_{\max}$ are temperature parameters at which the growth rate is zero, with finite growth occurring between $T_{\min}$

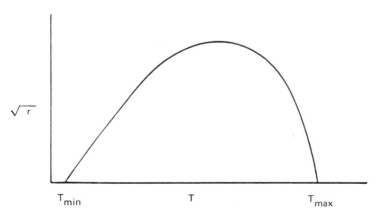

Figure 8.5

and T_{max}, and where β and γ are two further parameters that enable (8.2.3) to fit such data throughout the entire biokinetic temperature range. The statistical properties in estimation of the "square-root" model (8.2.3) are excellent.

A major part of the reason why (8.2.1) and (8.2.2) have poor statistical properties in estimation is that those models are overparameterized, having six parameters where, as (8.2.3) shows, four parameters are sufficient (at least for one large area of poikilotherm development). As was the case with the rational functions models described earlier in this section, use of unnecessary parameters can easily destroy the estimation properties of a model. Therefore, one should always strive to have only as many parameters in a model as are needed for that model to provide a good fit to the data.

8.3 OTHER EXCLUDED MODELS

A further group of models not included in Chapters 4–6 of this handbook are those rational functions models for which the degree of the polynomial in the numerator of the expression is greater than the degree of the polynomial in the denominator of the expression, for example,

$$y = (\beta + \gamma X + \delta X^2)/(1 + \alpha X) \tag{8.3.1}$$

Model (8.3.1) contrasts with (4.3.24), (6.4.11), and (6.5.1), in that those rational functions models have the degree of the denominator polynomial equal to, or greater than, the degree of the numerator polynomial. The advantage of the latter is that y approaches an asymptote as $X \to \infty$, whereas (8.3.1) lacks that property. I (unpublished results) have studied (8.3.1) in combination with many of the data sets for which (6.4.11) was also used (Ratkowsky, 1987) and have found that (8.3.1) fitted the data no better, on average, than (6.4.11), a model with the same number of parameters, and that it had poorer estimation properties than

(6.4.11). It appears that the best rational functions models to use are those for which the degree of the polynomial in the denominator is equal to, or greater than, the degree of the polynomial in the numerator.

9

Obtaining Good Initial Parameter Estimates

All methods for determining the least-squares estimates of the parameters in a nonlinear regression model require initial parameter estimates as the starting point of an iterative or search method. The Gauss–Newton method is favored in this handbook, as it converges rapidly for close-to-linear models. A linear model converges in a single step from any starting point (Ratkowsky, 1983, Appendix 2.A), and close-to-linear models require but a few steps, provided the initial estimates are "reasonable." This chapter is devoted to a discussion of how to achieve reasonable initial parameter estimates.

For models whose parameters are exclusively expected-value parameters, obtaining initial estimates is a very simple matter. As noted in Section 2.3.1 under "Finding Expected-Value Parameters" for the case of a single explanatory variable, one simply draws a smooth curve through the graph of the data for Y versus X

and, for the chosen values X_1, X_2, etc., of X, reads approximate y_1, y_2, etc., values from the graph. If the chosen X values correspond to ones that are in the data set, it is usually sufficient to use the *observed* Y values as the initial estimates. Because models whose parameters are all expected-value parameters always have good estimation properties (unless the X values fall outside the range of the observed data), convergence to the least-squares estimates is virtually assured and should occur in only a few iterations. For models that have one or more expected-value parameters but also contain other kinds of parameters, the initial estimates of the expected-value parameters are obtained in the same fashion, either by reading approximate fitted values from a graph or by using the observed Y values corresponding to the chosen X values.

A wide variety of methods for obtaining good initial parameter estimates was presented by Ratkowsky (1983), Chapter 8. One principle is to try to convert a model into a linear expression after guessing at the values of one or more parameters. When the parameters to be guessed are intercepts or asymptotes, a reasonable initial estimate is usually possible. For example, consider the four-parameter logistic model (5.4.1) (with the expected value y of Y on the left-hand side of the expression being replaced by Y itself, as no account is being taken of the stochastic assumption in this chapter),

$$Y = \delta + \alpha/[1 + \exp(\beta - \gamma X)]$$

From Figure 5.15, the lower asymptote is δ and the upper asymptote is $\alpha + \delta$. Hence, from data covering the whole range of the response (data not covering the whole range will result in poorly determined parameter estimates and far-from-linear behavior), one obtains an estimate δ_0 of δ by guessing the lower asymptote and an estimate α_0 of α from the difference between the guesses of the upper and lower asymptotes. Using α_0 and δ_0 in place of α and δ, respectively, and rearranging (5.4.1) and taking logarithms, one obtains

$$\log\{[\alpha_0/(Y - \delta_0)] - 1\} = \beta - \gamma X$$

for which the left-hand side, which can be evaluated for each of the data points (unless $Y \leq \delta_0$ or $Y \geq \alpha_0 + \delta_0$, in which case those points should not be used in determining the initial estimates), can be regressed on X to obtain estimates β_0 and γ_0 of β and γ, respectively, by simple linear regression.

A similar strategy can be applied to the four-parameter Gompertz curve (5.4.9),

$$Y = \delta + \alpha \exp[-\exp(\beta - \gamma X)]$$

As was the case with the logistic model (5.4.1), estimates α_0 of α and δ_0 of δ can be obtained from guesses of the upper and lower asymptotes. Rearrangement, followed by taking logarithms twice, leads to

$$\log\{-\log[Y - \delta_0)/\alpha_0]\} = \beta - \gamma X$$

for which the left-hand side, which can be evaluated for each of the data points (unless $Y \leq \delta_0$ or $Y \geq \alpha_0 + \delta_0$, in which case those points should not be used in determining the initial estimates), can be regressed on X to obtain estimates β_0 and γ_0 of β and γ, respectively, using simple linear regression.

Another class of models for which it is easy to obtain good initial estimates are the rational functions models. The most complicated one considered in this handbook is (6.5.1),

$$Y = (\beta + \gamma X + \epsilon X^2)/(1 + \alpha X + \delta X^2)$$

This equation may be rearranged and written as

$$Y = \beta + \gamma X + \epsilon X^2 - \alpha XY - \delta X^2 Y$$

One can now obtain initial estimates of the five parameters α, β, γ, δ, and ϵ from the multiple linear regression (with an intercept) of Y on X, X^2, XY, and $X^2 Y$. The initial estimates so obtained should converge readily to the least-squares estimates, provided (6.5.1) really does fit the data well.

Other strategies have to be adopted for a model such as (4.3.34),

$$Y = \exp[\alpha + \beta/(X + \gamma)]$$

Taking the logarithm of both sides yields

$$\log Y = \alpha + \beta/(X + \gamma)$$

which is linear in α and β but not in γ. One can take advantage of the fact that α and β can be estimated by simple linear regression of $\log Y$ on $1/(X + \gamma_0)$, where γ_0 is a constant value of γ. By varying γ_0 systematically over a wide range of values and estimating corresponding values of α_0 and β_0 by linear regression, one can find the set of α_0, β_0, γ_0 values that approximately minimizes the sum of squares in Y (or $\log Y$). This set of values can then serve as initial values for obtaining the least-squares estimates. This strategy of changing a parameter such as γ in (4.3.34) systematically over a wide range of values, often in equal increments, enabling the remaining parameters to be estimated by ordinary least squares, was widely employed previously (Ratkowsky, 1983, Chap. 8).

Certain models pose more difficult problems. Consider the sum of two exponential terms, (4.4.4),

$$Y = \alpha \exp(-\beta X) + \gamma \exp(-\delta X)$$

That parameterization had poor estimation properties, so that parameters α and γ were replaced by expected-value parameters y_1 and y_2, respectively. Estimates of y_1 and y_2 are readily obtained in the usual way for expected-value parameters by reading the approximate fitted values of Y from a graph of Y versus X, but estimates of β and δ, or of $\lambda_1 = \exp(-\beta)$ and $\lambda_2 = \exp(-\delta)$, which have slightly better properties than β and δ, have to be obtained by other means. The traditional method in pharmacology is to plot $\log Y$ versus X, from which a graph such as Figure 9.1 may be obtained. If a sum of exponentials models is appropriate, the graph should be linear at high values of X, as shown. The absolute value of the slope of the straight line is an estimate of β, which can be converted to an estimate of λ_1. For the remaining data points that are not in the straight-line region, the fitted values of $\log Y$ are subtracted from the observed $\log Y$ for each X value, creating adjusted values of $\log Y$ for the data points not in the straight-line region. This limited set of adjusted $\log Y$ values is plotted against the corresponding X values and, if a sum of two exponentials models is in fact the true model, the graph should approximate a straight line, from which an estimate of δ is given by the absolute value of the slope, which can be converted to an estimate of λ_2. Thus, estimates of y_1, y_2, λ_1, and λ_2 have been obtained for use with (4.4.6).

Obtaining good initial estimates is an important feature of nonlinear regression modeling. Far-from-linear models are unlikely to converge unless the initial estimates are close to the least-squares estimates. That is, the "zone of convergibility" of a far-from-linear model is much less than the zone of convergibility of a close-to-linear model. As already mentioned at the beginning of this

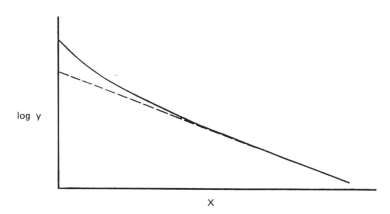

log y

X

Figure 9.1

section, the Gauss–Newton method will converge in a single step from *any* set of initial estimates, no matter how far removed they may be from the least-squares estimates, if the model is a linear model. Although nonlinear regression models require more than a single step, it should be clear that the closer a model is to behaving like a linear model, the more likely it is to converge in a small number of iterations from a reasonable starting point and, more important, the zone of convergibility is greater for a close-to-linear model than a far-from-linear one. The emphasis in this handbook has been on the identification of models that are close-to-linear in their estimation behavior. The closer-to-linear the model, the less care one needs for obtaining initial parameter estimates. Nevertheless, some care is required because, although the zone of convergibility may be relatively large, there will be a region outside that zone for which convergence will not occur.

10

Summary

It has been the purpose of this handbook to identify and present close-to-linear nonlinear regression models for practical use. The quest for good parameterizations was the central aspect of the models presented in Chapters 4–8. This concluding chapter attempts to summarize some of the noteworthy features of nonlinear regression modeling in the hope of highlighting some aspects of the subject that might otherwise be overlooked. The overall goal is to impart a unified basis to the search for models with good estimation properties.

10.1 BACKGROUND TO THE QUEST FOR CLOSE-TO-LINEAR MODELS

Nonlinear regression models differ from linear regression models in two important ways, as shown by Beale (1960) and Bates and

Watts (1980). One major difference is that nonlinear regression models have a *curved*, rather than a *straight*, solution locus, which is the set of all possible values the parameters may assume, the least-squares solution being the one that is closest to the observed vector of Y values. The consequences of this curved solution locus, the extent of the curvature being a measure of the intrinsic nonlinearity, is that the fitted Y values and the residual sum of squares are biased. However, the extent of this bias is generally trivial, as most models in practical use have very small intrinsic nonlinearities. The second major way in which nonlinear regression models differ from linear regression models has to do with the particular parameterization of the nonlinear regression model, from which the name "parameter effects" nonlinearity derives (Bates and Watts, 1980). This aspect of nonlinearity is a consequence of the lack of uniformity of the coordinate system on the solution locus (or the tangent plane to the solution locus). For a linear model, the parameter lines are straight and parallel; for a nonlinear model, they are neither straight nor parallel, the greater the nonlinearity, the greater the curvature of the lines and the departure from parallelism. Changing the parameterization results in a new coordinate system on the solution locus that corresponds to the new set of parameters. The closer-to-linear the parameterization, the straighter and more parallel is the coordinate system corresponding to that parameterization. If the solution locus is significantly curved, no amount of searching for reparameterizations can alter the fact that, with even the best possible parameterization, the final model must still exhibit significant nonlinear behavior. Fortunately, as already reiterated, very few models in practical use have a significant intrinsic nonlinearity, so that the search for better parameterizations does have a purposeful basis.

Expected-value parameters (see Sec. 2.3.1) seem to be the best kind of parameters to have in one's model. Since expected-value parameters are nothing more than fitted values of the model (for given specified X values), their biases are determined only by the intrinsic nonlinearity of the model which, as we have seen, is generally small. Because the extent of the bias in an estimator of a parameter is closely associated with the extent of nonnormal behavior, the amount of nonnormality, and hence nonlinear behavior, in an expected-value parameter is small.

One limitation of expected-value parameters is that their defining equations often cannot be solved for the "old" parameters in terms of the expected-value parameters, so that one cannot always obtain an algebraic expression for the reparameterized model. However, as shown by Dr. I.W. Saunders (unpublished results), one can always, in principle, solve *numerically*, rather than *algebraically*, the least-squares equations for the parameterization containing only expected-value parameters. Also, one can calculate the measures of nonlinear behavior, such as the Bates and Watts (1980) measures and the Hougaard (1985) skewness measure, for the new parameterizations. Nevertheless, the solution requires good initial estimates of the parameters in the *original* formulation for it to converge, so that this is a practical disadvantage in using a model with expected-value parameters that have to be evaluated numerically.

This handbook has concentrated on the presentation of close-to-linear models. The benefits of close-to-linear models are several:

1. The least-squares estimates can be obtained in a straightforward fashion from a set of reasonable initial estimates.
2. The estimators are close to being unbiased, normally distributed, minimum variance estimators in small samples as well as large.
3. There is a sound basis for making comparisons between parameters in more than one set of data. Such comparisons are meaningful and are readily carried out.
4. Predicted values of the response variable will be almost unbiased, as will the residual sum of squares.
5. Joint confidence regions for the parameters will be close to being "ellipsoidal," and confidence limits for individual parameters will be close to being symmetrical.
6. In general, the properties of a close-to-linear nonlinear regression model will be similar to those of a linear regression model (hence the name "close-to-linear"). It is generally

recognized that the least-squares estimators of the parameters in linear regression models have optimum properties. Hence, a close-to-linear nonlinear regression model is as close as one can get, with a nonlinear regression model, to achieving optimality.

10.2 SOME RULES OR GUIDELINES

Readers have already been confronted with Ockham's Razor (Sec. 2.5.4), which states that "plurality should not be assumed without necessity." This philosophical principle provides a sound basis for nonlinear regression modeling because, in general, the more complex the model, the greater is the extent of nonlinear behavior exhibited by that model. It is possible to develop a few rules or guidelines that, although not theorems in the mathematical sense, are principles that experience has suggested are generally true.

Rule 1: Models with few parameters are more likely to be close-to-linear than models with many parameters. For example, most one-parameter models, such as the ones in Sections 4.1, 5.1, and 6.1, are close-to-linear in their original parameterizations. That is not to say that they cannot be made even *closer-to-linear* by reparameterization, especially by replacing the parameters with expected-value parameters. Indeed, all throughout Section 4.1, the original parameterizations, being close-to-linear, as indicated by a single asterisk ($*$), are improved upon by the use of an expected-value parameterization, to produce a model that is even closer-to-linear, as indicated by a double asterisk ($**$).

Rule 2: If one model is a special or restricted case of another model, it will almost certainly display less nonlinear behavior than the other model. For example, compare (4.1.1),

$$y = \log(X - \alpha)$$

with (4.2.1),

$$y = \beta \log(X - \alpha)$$

The former is a special case of the latter, with $\beta = 1$, and will be closer-to-linear than the latter. So too will

$$y = 2.51 \log(X - \alpha)$$

where β has now been arbitrarily assigned the constant value 2.51. That is, provided β assumes some fixed value, not to be estimated, the behavior of the resulting model will be better than the original. Of course, if α is a constant in (4.2.1), the resulting model is a linear model.

Consider now the four-parameter curves having a maximum or a minimum that are given by (6.4.4–6.4.8). Each of these models contains the term $X - \delta$, where δ functions to allow a maximum or a minimum to occur at the value $X = \delta$ rather than at $X = 0$. In each case, letting $\delta = 0$ reduces the model in question to a three-parameter model that is closer to behaving like a linear model than the original four-parameter model containing a nonzero δ. Thus, each of the following models is closer-to-linear than its four-parameter counterpart.

Bragg (or normal-type):

$$(*) \quad y = \alpha + \beta \exp(-\gamma X^2) \tag{10.2.1}$$

Lorentz:

$$(*) \quad y = \alpha + \beta/(1 + \gamma X^2) \tag{10.2.2}$$

Bacon:

$$(*) \quad y = \alpha + \beta(\cos X)^{\gamma} \tag{10.2.3}$$

Ergun:

$$(*) \quad y = \alpha + \beta \exp(-\gamma \sin^2 X) \tag{10.2.4}$$

Ruland:

$$(*) \quad y = \alpha + \beta/(1 + \gamma \sin^2 X) \tag{10.2.5}$$

Rule 3: A multiparameter model that exhibits close-to-linear behavior will not have its behavior improved significantly by forcing one of its parameters to take on a fixed value, as in Rule 2. This is really just another way of saying that if a model already exhibits close-to-linear behavior, there is little latitude for improvement. For example, the Holliday model (4.3.18) exhibited so little nonlinear behavior in combination with data on the yield of onions that when γ was set equal to zero to produce the "asymptotic" or "reciprocal" model (4.2.20), little improvement occurred (Ratkowsky, 1983, Table 3.1).

Rule 4: Even a model with poor statistical behavior can become close-to-linear if the sample size is large enough. In Section 2.1, a close-to-linear nonlinear regression model was defined as one whose parameter estimators came close to achieving the "asymptotic" (that is, large sample) properties of being unbiased, normally distributed, minimum variance estimators. By making the sample size progressively larger, the estimation properties of any nonlinear regression model progressively improves. Thus, even a far-from-linear model can become a close-to-linear model if a sufficiently large sample of data is used.

Were it not for this fact, it would not be practical to attempt to model complex systems with nonlinear regression models, as the properties of the estimators of the parameters in such complex systems would almost certainly be far-from-linear, rendering use of the parameter estimates uncertain because of their large biases. By employing a sufficiently large sample size, it is possible to overcome this problem.

For example, consider a model for total tuber yield for the potato cultivar Kennebec grown on Krasnozem soils in Tasmania (Thorp and Wallace, 1987). The model concerned the response (total tuber yield) to the application of nitrogen (N), phosphorus (P), and potassium (K) fertilizer. The potatoes were grown at 27 sites, with the NPK fertilizer trials each having either three or four levels for each factor, so that, at each site, a 3^3 or 4^3 factorial design was used. This produced a set of 1345 data points for use in the model. Each site had a particular paddock history and pH. A nitrogen rating factor N_r, having three classes of paddock fertility based on cropping history, was assigned to each site, since soil nitrogen could not be determined directly. For P and K, a bicarbonate test was used, and site mean bicarbonate soil test results, denoted P_t and K_t, were determined. Since the response of a crop to fertilizer has been generally found by agricultural scientists to follow the Mitscherlich law [that is, the asymptotic regression equation (4.3.1–4.3.9)], the following model was proposed by Thorp and Wallace (1987):

$$y = A\{1 - \exp[-(\alpha N + N_r)]\}\{1 - \exp[-(\beta P + \gamma P_t)]\}$$
$$\times \{1 - \exp[-(\delta K + \nu K_t)]\} \quad (10.2.6)$$

where the response variable y is predicted yield, and N, P, and K are the amounts of applied fertilizer in each factorial experiment. The additional explanatory variables are, respectively, N_r, the three-level nitrogen fertility factor of the paddock; P_t, the site mean bicarbonate soil phosphorus test result; and K_t, the site mean bicarbonate soil potassium test result. To simplify the notation in (10.2.6), subscripts have been omitted but, although there is a single value over the whole model for the parameters α, β, γ, δ, and ν and for each of the three factor levels of N_r, there is a separate asymptote A (maximum site yield) for each of the 27 sites. Thus,

35 coefficients were determined using (10.2.6) in combination with the 1345 data points, for 1310 degrees of freedom, which is a very substantial effective sample size.

The large sample size results in the behavior of (10.2.6) being quite close-to-linear. Although it would be impractical to calculate the various measures of nonlinear behavior in Section 2.2 for such a large data set (computer time and memory requirements would be considerable), one can obtain clues to the behavior in estimation of (10.2.6) by looking at the t values, the ratios of the parameter estimates to the standard errors (SE) of the estimates. These are shown in Table 10.1 for the above-described data set.

Although t values have limitations as measures of nonlinear behavior because they are not location-independent (see Ratkowsky, 1983, p.201), their large values for the model/data set combination under consideration suggest that each of the parameters is very well determined. Only the calculation of the nonlinearity measures or the carrying out of a simulation study can settle the question of really how close-to-linear the behavior of the es-

Table 10.1   Estimated Parameters for Model (10.2.6) for Kennebec Variety Potatoes Grown on Krasnozem Soils

Parameter	Estimate	SE	t value
A	82.0	1.85	44.2
α	1.06	0.0854	12.4
N_{r1}	0.818	0.0141	58.1
N_{r2}	0.863	0.0163	53.1
N_{r3}	1.19	0.0176	67.3
β	13.7	0.516	26.6
γ	24.8	0.552	44.9
δ	5.34	0.526	10.2
ν	9.99	0.289	34.6

Note: The values for the asymptote A were averaged over the 27 sites.

timators is in this problem. The rather small SE values (or, equivalently, large t values) give one grounds for the belief that the very large sample size has overcome the potentially high non-linear behavior that would have been observed for a model of this complexity if a smaller data set had been used.

Rule 5: Reducing the residual variance will reduce the extent of nonlinearity. The various measures of nonlinear behavior are influenced by the residual variance such that by reducing the residual variance, the behavior of the parameter estimators is improved. This follows from the fact that the bias in the estimator is proportional to the residual variance (Box, 1971), and the Bates and Watts (1980) measures of intrinsic and parameter-effects nonlinearities are proportional to the square root of the residual variance. Thus, reducing the residual variance by more careful experimentation or by more careful selection of the experimental material can have beneficial results. Thus, there are two ways of making the estimators approach their asymptotic behavior more closely, that is, (1) by increasing the sample size (Rule 4), and (2) decreasing the residual variance (Rule 5).

Although the aim of this handbook has been to try to find which parameterization is the closest-to-linear for each basic model, it is likely that there will be sets of data for which the best parameterization is different from the one indicated here. This will certainly be true for some data sets for which the observations are restricted to only a portion of the potential range of response of the model. Thus, if a model is capable of describing an asymptote but if there are no data points near the asymptote, one cannot expect the model/data set combination to exhibit good estimation behavior. Fragmentary data are therefore best avoided. The behavior of the models in this handbook was assessed using an arbitrary set of parameters, with random normal error being added to the generated response value for each X value. In all cases, the data were required to cover a wide response range. There is, however, no guarantee that the parameter considered in this handbook as the worst-behaving one in the model in question and hence most in need of reparameterization will in fact be the worst-behaving one for *all* real data sets. Thus, a particular combination of parameter values for models in some disciplines could result in some other parameter being the worst-behaving one.

One way of seeing in how close-to-linear a manner a model/data set combination behaves without calculating the various nonlinearity measures in Sections 2.2.1 and 2.2.2 is to carry out a simulation study (see Sec. 2.2.3). A modeler who frequently uses a particular model will find it worthwhile to conduct a few simulation studies on some typical data sets. This will reveal whether the guidelines given in this handbook about which parameters need reparameterization are adequate or whether a different parameterization is needed for that particular problem. For other modelers, however, who use a wide range of models with only a few sets of data per model, it is hoped that the models marked with one asterisk (*) or two asterisks (**) in this handbook do indeed come close to exhibiting the behavior that one usually can expect only from a linear model.

References

Amari, S.I. (1982). Differential Geometry of Curved Exponential Families—Curvatures and Information Loss, *Ann. Statist. 10*, 375–385.

Bacon, D.W. and Watts, D.G. (1971). Estimating the Transition Between Two Intersecting Straight Lines, *Biometrika 58*, 525–534.

Bacon, G.E. (1956). A Method for Determining the Degree of Orientation of Graphite, *J. Applied Chem. 6*, 447–481.

Bard, Y. (1974). *Nonlinear Parameter Estimation*, Academic Press, New York.

Bates, D.M. and Watts, D.G. (1980). Relative Curvature Measures of Nonlinearity, *J.R. Statist. Soc., Ser. B 42*, 1–25.

Bates, D.M. and Watts, D.G. (1988). *Nonlinear Regression Analysis and Its Applications*, Wiley, New York.

Beale, E.M.L. (1960). Confidence Regions in Nonlinear Estimation, *J.R. Statist. Soc., Ser. B 22*, 41–76.

Bělehrádek, J. (1935). *Temperature and Living Matter*, Protoplasma Monograph 8, Borntraeger, Berlin.

Bellman, R.E. (1957). *Dynamic Programming*, Princeton University Press, Princeton, N.J.

Belsley, D.A., Kuh, E., and Welsch, R.E. (1980). *Regression Diagnostics: Identifying Influential Data and Sources of Variation*, Wiley, New York.

Bleasdale, J.K.A. and Nelder, J.A. (1960). Plant Population and Crop Yield, *Nature 188*, 342.

Box, M.J. (1971). Bias in Nonlinear Estimation, *J.R. Statist. Soc., Ser. B33*, 171–201.

Bradley, R.S. (1936). Polymolecular Adsorbed Films. I. The Adsorption of Argon on Salt Crystals at Low Temperatures, and the Determination of Surface Yields, *J. Chem. Soc. 1936*, 1467–1474.

Bragg, R.H. and Packer, C.M. (1962). Orientation Dependence of Structure in Pyrolytic Graphite, *Nature, London 195*, 1080–1082.

Brunauer, S., Emmett, P.H., and Teller, E. (1938). Adsorption of Gases in Multimolecular Layers, *J. Am. Chem. Soc. 60*, 309–319.

Campbell, N.A. and Keay, J. (1970). Flexible Techniques in Describing Mathematically a Range of Response Curves of Pasture Species. Proceedings of the Eleventh International Grasslands Congress, Surfers Paradise, Queensland, Australia, pp. 332–334.

Causton, D.R. and Venus, J.C. (1981). *The Biometry of Plant Growth*, Arnold, London.

Chambers, J. (1973). Fitting Nonlinear Models: Numerical Techniques, *Biometrika 60*, 1–13.

Chen, C.S. (1971). Equilibrium Moisture Curves for Biological Materials, *Trans. ASAE (Am. Soc. Agric. Eng.) 14*, 924–926.

Chirife, J. and Iglesias, H.A. (1978). Equations for Fitting Water Sorption Isotherms of Foods: Part 1—A Review, *J. Food Technol. 13*, 159–174.

Clarke, G.P.Y. (1980). Moments of the Least Squares Estimators in a Nonlinear Regression Model. *J.R. Statist. Soc., Ser. B 42*, 227–237.

Cook, R.D. and Goldberg, M.L. (1986). Curvatures for Parameter Subsets in Nonlinear Regression, *Ann. Statist. 14*, 1399–1418.

Cook, R.D. and Weisberg, S. (1982). *Residuals and Influence in Regression*, Chapman and Hall, London.

Cook, R.D. and Witmer, J.A. (1985). A Note on Parameter-Effects Curvature, *J. Amer. Statist. Assoc. 80*, 872–878.

Cook, R.D., Tsai, C.L., and Wei, B.C. (1986). Bias in Nonlinear Regression, *Biometrika 73*, 615–623.

Cox, D.R. and Snell, E.J. (1968). A General Definition of Residuals, *J.R. Statist. Soc., Ser. B 30*, 248–275.

D'Arcy, R. L. and Watt, I. C. (1970). Analysis of Sorption Isotherms of non-Homogeneous Sorbents, *Trans. Faraday Soc. 66*. 1236–1245.

Dielman, T. (1983). Pooled Cross-sectional and Time Series Data: a Survey of Current Statistical Methodology, *American Statistician 37*, 111–122.

Donaldson, J.R. and Schnabel, R.B. (1987). Computational Experience with Confidence Regions and Confidence Intervals for Nonlinear Least Squares, *Technometrics 29*, 67–82.

Draper, N.R. (1984). The Box-Wetz Criterion versus R^2, *J.R. Statist. Soc. A 147*, 100–103.

Draper, N.R. and Smith, H. (1981). *Applied Regression Analysis*, 2nd. ed., Wiley, New York.

Ergun, S. (1976). Analysis of Coherence, Strain, Thermal Vibration and Preferred Orientation in Carbons by X-ray Diffraction, *Carbon 14*, 139–150.

Fabens, A.J. (1965). Properties and Fitting of the von Bertalanffy Growth Curve, *Growth 29*, 265–289.

Farazdaghi, H. and Harris, P.M. (1968). Plant Competition and Crop Yield, *Nature 217*, 289–290.

Francis, R. I. C. C. (1988a). Maximum Likelihood Estimation of Growth and Growth Variability from Tagging Data, *New Zealand J. Mar. Freshwater Res. 22*, 42–51.

Francis, R. I. C. C. (1988b). Are Growth Parameters Estimated from Tagging and Age-length Data Comparable? *Can. J. Fish. Aquat. Sci. 45*, 936–942.

Freundlich, H. (1926). *Colloid and Capillary Chemistry*, Methuen, London.

Gallant, A. R. (1976). Confidence Regions for the Parameters of a Nonlinear Regression Model, *Institute of Statistics Mimeograph Series No. 1077*, The Institute of Statistics, University of North Carolina, Raleigh, NC.

Gallant, A.R. (1987). *Nonlinear Statistical Models*, Wiley, New York.

Gillis, P.R. and Ratkowsky, D.A. (1978). The Behaviour of Estimators of the Parameters of Various Yield-Density Relationships, *Biometrics 34*, 191–198.

GLIM (Generalised Linear Interactive Modelling) (1978). Numerical Algorithms Group, Oxford, U.K.

Gnanadesikan, R. (1977). *Methods for Statistical Data Analysis of Multivariate Observations*, Wiley, New York.

Griffiths, D.A. and Miller, A.J. (1973). Hyperbolic Regression—A Model Based on Two-Phase Piecewise Linear Regression with a Smooth Transition between Regimes, *Commun. Statist. 2*, 561–569.

Gunary, D. (1970). A New Adsorption Isotherm for Phosphate in Soil, *J. Soil Sci. 21*, 72–77.

Gunst, R.F. and Mason, R.L. (1980). *Regression Analysis and Its Application*, Marcel Dekker, New York.

Hailwood, A.J. and Horrobin, S. (1946). Absorption of Water by Polymers: Analysis in Terms of a Simple Model, *Trans. Farad. Soc. 42B*, 84–102.

Halperin, M. (1963). Confidence Interval Estimation in Non-linear Regression, *J. R. Statist. Soc., Ser. B25*, 330–333.

Halsey, G. (1948). Physical Adsorption on Non-uniform Surfaces, *J. Chem. Phys. 16*, 931–937.

Hamilton, D. (1986). Confidence Regions for Parameter Subsets in Nonlinear Regression, *Biometrika 73*, 57–64.

Hartley, H. O. (1964). Exact Confidence Regions for the Parameters in Non-linear Regression Laws, *Biometrika 51*, 347–353.

Hastings, C. Jr. (1955). *Approximating for Digital Computers*, Princeton University Press, Princeton, N.J.

Healy, M.J.R. (1984). The Use of R^2 as a Measure of Goodness of Fit, *J.R. Statist. Soc. A 147*, 608–609.

Helland, I.S. (1987). On the Interpretation and Use of R^2 in Regression Analysis, *Biometrics 43*, 61–69.

Henderson, S.M. (1952). A Basic Concept of Equilibrium Moisture, *Agric. Eng. 33*, 29–32.

Hoaglin, D.C., Mosteller, F., and Tukey, J.W. (eds.) (1983). *Understanding Robust and Exploratory Data Analysis*, Wiley, New York.

Holliday, R. (1960). Plant Population and Crop Yield, *Field Crop Abstr. 13*, 159–167, 247–254.

Horton, W.S. (1979). An Examination of Five Preferred Orientation Functions, *Carbon 17*, 153–155.

Hougaard, P. (1985). The Appropriateness of the Asymptotic Distribution in a Nonlinear Regression Model in Relation to Curvature, *J.R. Statist. Soc., Ser. B 47*, 103–114.

Hougen, O.A. and Watson, K.M. (1947). *Chemical Process Principles*, Part III, Wiley, New York.

Iglesias, H.A. and Chirife, J. (1981). An Equation for Fitting Uncommon Water Sorption Isotherms in Food. *Lebens. Wiss. Technol. 14(2)*, 105–106.

Iglesias, H.A. and Chirife, J. (1982). *Handbook of Food Isotherms*, Academic Press, New York.

Jolicoeur, P. (1975). Linear Regressions in Fishery Research: Some Comments, *J. Fish. Res. Bd. Can. 32*, 1491–1494.

Kinsella, J.E. and Fox, P.F. (1986). Water Sorption by Proteins: milk and whey proteins, *CRC Critical Reviews in Food Science and Nutrition 24(2)*, 91–139.

Kühn, I. (1967). Generalized Potential Theory of Adsorption. I. Derivation of a General Equation for Adsorption Isotherms, *J. Colloid Interface Sci. 23*, 563–571.

Langmuir, I. (1916). Constitution and Fundamental Properties of Solids and Liquids. I. Solids, *J. Am. Chem. Soc. 38*, 2221–2295.

Lawless, J.F. (1982). *Statistical Models and Methods for Lifetime Data*, Wiley, New York.

Lowry, R.K. and Morton, R. (1983). An Asymmetry Measure for Estimators in Non-linear Regression Models, *Proc. 44th Session Int. Statist. Inst., Madrid, Contributed Papers, Vol. 1*, 351–354.

Lowry, R.K. and Ratkowsky, D.A. (1983) A Note on Models for Poikilotherm Development, *J. Theor. Biol. 105*, 453–459.

Mead, R. (1979). Competition Experiments, *Biometrics 35*, 41–54.

Meyerhoff, M.E. and Robins, R.H. (1980). Disposable Potentiometric Ammonia Gas Sensors for Estimation of Ammonia in Blood, *Anal. Chem. 52*, 2383–2387.

Miller, A.J. (1984). Selection of Subsets of Regression Variables (with discussion), *J. R. Statist. Soc., A147*, 389–425.

Morgan, P.H., Mercer, L.P., and Flodin, N.W. (1975). General Model for Nutritional Responses of Higher Organisms, *Proc. Nat. Acad. Sci. USA 72*, 4327–4331.

Morton, R. (1987). Asymmetry of Estimators in Nonlinear Regression, *Biometrika 74*, 679–685.

Nelder, J.A. and Wedderburn, R.W.M. (1972). Generalised Linear Models. *J.R. Statist. Soc., Ser. A135*, 370–384.

Oswin, C.R. (1946). Kinetics of Package Life. III. Isotherm, *J. Soc. Chem. Ind. 65*, 419–421.

Prunty, L. (1983). Curve Fitting with Smooth Functions That Are Piecewise-linear in the Limit, *Biometrics 39*, 857–866.

Ratkowsky, D.A. (1983). *Nonlinear Regression Modeling: a Unified Practical Approach*, Marcel Dekker, New York.

Ratkowsky, D.A. (1985). A Statistically Suitable General Formulation for Modelling Catalytic Chemical Reactions, *Chem. Eng. Sci. 40*, 1623–1628.

Ratkowsky, D.A. (1986a). A Suitable Parameterization of the Michaelis–Menten Enzyme Reaction, *Biochem. J. 240*, 357–360.

Ratkowsky, D.A. (1986b). A Statistical Study of Seven Curves for Describing the Sorption of Phosphate by Soil, *J. Soil Sci. 37*, 183–189.

Ratkowsky, D.A. (1986c). Statistical Properties of Alternative Parameterizations of the van Bertalanffy Growth Curve, *Can. J. Fish. Aquat. Sci. 43*, 742–747.

Ratkowsky, D.A. (1986d). A Statistical Examination of Five Models for Preferred Orientation in Carbon Materials, *Carbon 24*, 211–215.

Ratkowsky, D.A. (1987). The Statistical Properties and Utility of Rational Functions Models, *Can. J. Chem. Eng. 65*, 845–851.

Ratkowsky, D.A., Lowry, R.K., McMeekin, T.A., Stokes, A.N., and Chandler, R.E. (1983). Model for Bacterial Culture Growth Rate throughout the Entire Biokinetic Temperature Range, *J. Bacteriol. 154*, 1222–1226.

Ratkowsky, D.A., Olley, J., McMeekin, T.A., and Ball, A. (1982). Relationship between Temperature and Growth Rate of Bacterial Cultures, *J. Bacteriol. 149*, 1–5.

Ratkowsky, D.A., and Reedy, T.J. (1986). Choosing Near-linear Parameters in the Four-parameter Logistic Model for Radioligand and Related Assays, *Biometrics 42*, 575–583.

Reid, D. (1978). The Effects of Frequency of Defoliation on the Yield Response of a Perennial Ryegrass Sward to a Wide Range of Nitrogen Application Rates, *J. Agr. Sci. (Cambridge) 90*, 447–457.

Richards, F.J. (1959). A Flexible Growth Function for Empirical Use, *J. Exp. Biol. 10*, 290–300.

Ricker, W.E. (1973). Linear Regressions in Fishery Research, *J. Fish. Res. Bd. Can. 30*, 409–434.

Ross, G.J.S. (1970). The Efficient Use of Function Minimization in Non-linear Maximum-likelihood Estimation, *J.R. Statist. Soc.*, *Ser. C 19*, 205–221.

Ross, G.J.S. (1975). Simple Non-linear Modelling for the General User, *Proc. 40th Session Int. Statist. Inst.*, *Warsaw, Contributed Papers, Vol. 2*, 585–593.

Ross, G.J.S. (1978). Exact and Approximate Confidence Regions for Functions of Parameters in Non-Linear Models. *COMPSTAT 78*, Third Symposium on Computation, Physica-Verlag, Vienna.

Ruland, W. (1967). X-ray Studies on Preferred Orientation in Carbon Fibers, *J. Appl. Phys. 38*, 3585–3589.

Schmidt, R. (1982). *Advances in Nonlinear Parameter Optimization*, Springer-Verlag, Berlin.

Schnute, J. and Fournier, D. (1980). A New Approach to Length-Frequency Analysis: Growth Structure, *Can. J. Fish. Aquat. Sci. 37*, 1337–1351.

Schoolfield, R.M., Sharpe, P.J.H., and Magnuson, C.E. (1981). Non-linear Regression of Biological Temperature-dependent Rate Models Based on Absolute Reaction-rate Theory, *J. Theor. Biol. 88*, 719–731.

Sharpe, P.J.H. and De Michele, D.W. (1977). Reaction Kinetics of Poikilotherm Development, *J. Theoret. Biol. 64*, 649–670.

Shinozaki, K. and Kira, T. (1956). Intraspecific Competition Among Higher Plants, VII: Logistic Theory of the C-D Effect, *J. Inst. Polytech.*, *Osaka City Univ.*, *Ser. D7*, 35–72.

Sibbesen, E. (1981). Some New Equations to Describe Phosphate Sorption by Soils, *J. Soil Sci. 32*, 67–74.

Smith, S.E. (1947). The Sorption of Water Vapor by High Polymers, *J. Am. Chem. Soc. 69*, 646–651.

Sprent, P. and Dolby, G.R. (1980). Response to Query on the Geometric Mean Functional Relationship, *Biometrics 36*, 547–550.

Thorp, J.R. and Wallace, S.P. (1987). A Fertiliser Model for Predicting Potato Yields Based on Soil Tests, Proc. 4th Aust. Agronomy Conf., Australian Society of Agronomy, LaTrobe Univ., Melbourne, Aug. 24–27, 1987.

Tukey, J.W. (1977). Exploratory Data Analysis, Addison-Wesley, Reading, MA.

van der Berg, C. and Bruin, S. (1981). Water Activity and its Estimation in Food Systems: Theoretical Aspects, in *Water Activity: Influence on Food Quality*, L.B. Rockland and G.F. Stewart (eds.), Academic Press, New York, pp.1–55.

Watt, K.E.F. (1968). *Ecology and Resource Management*, McGraw-Hill, New York.

West, P.W. (1980). Use of Diameter Increment and Basal Area Increment in Tree Growth Studies, *Can. J. For. Res. 10*, 71–77.

West, P.W., Davis, A.W., and Ratkowsky, D.A. (1986). Approaches to Regression Analysis with Multiple Measurements from Individual Sampling Units, *J. Statist. Comput. Simul. 26*, 149–175.

West, P.W., Ratkowsky, D.A., and Davis, A.W. (1984). Problems of Hypothesis Testing of Regressions Analysis with Multiple Measurements from Individual Sampling Units, *Forest Ecol. Management 7*, 207–224.

Wilk, M.B. and Gnanadesikan, R. (1968). Probability Plotting Methods for the Analysis of Data, *Biometrics 55*, 1–17.

Williams, E. J. (1962). Exact Fiducial Limits in Non-linear Estimation, *J. R. Statist. Soc., Ser. B24*, 125–139.

Yang, R.C., Kozak, A., and Smith, J.H.G. (1978). The Potential of Weibull-type Functions as Flexible Growth Curves, *Can. J. For. Res. 8*, 424–431.

Table of Symbols

This is an incomplete list, involving symbols that tend to appear in two or more chapters of this handbook. Some symbols, such as $J(\theta^*)$ or $P(\theta^*)$, which appear in a single chapter or a single section of one chapter, may not be listed here.

A, a, B, b, C, c, etc.	Parameters (coefficients to be estimated)
exp	Exponential operator (base of natural logarithms e)
f	Indicates a mathematical function
log	Logarithmic operator; in this handbook, always signifying natural (Naperian) logarithms

n	Number of data points in sample
p	Number of parameters in regression model
tanh	Hyperbolic tangent, $\tanh(z) = [\exp(z) - \exp(-z)]/[\exp(z) + \exp(-z)]$.
X	Explanatory variable
$\mathbf{X}$	Vector of explanatory variables
X_1, X_2, X_3 (see also below)	Arbitrarily chosen values of the explanatory variable for which the fitted (expected) values of the response variable may be used as expected-value parameters (see Sec. 2.3.1)
X_1, X_2, X_3, X_4	Used to mean the first, second, third, and fourth explanatory variables, respectively (Chap. 7 only)
Y	Response variable
Y_t	Response variable at the tth data point
y	Expected value of the response variable
y_t	Expected value of the response variable at the tth data point
y_1, y_2, y_3, etc.	Expected-value parameters corresponding to arbitrarily chosen values of the explanatory variable X_1, X_2, X_3, etc.

$\alpha, \beta, \gamma, \delta$, etc.	Parameters (coefficients to be estimated)
Σ	Summation operator
σ^2	Error variance
*	Indicates a moderately close-to-linear model
**	Indicates a very close-to-linear model
***	Indicates a linear model

Appendix

This appendix supplements the Appendix to Ratkowsky (1983), pp. 209–270, by supplying the subroutine SKEW to calculate the Hougaard (1985) measure of skewness for each parameter of a nonlinear regression model (Sec. 2.2.2). Usage of SKEW is facilitated by inclusion of an illustrative example, the model function and data set being the same as that in the Example Programs section of Ratkowsky (1983), p. 236.

In the following pages, there will be found (1) the full FOR-TRAN listing of subroutine SKEW, (2) a FORTRAN calling program for fitting one of the parameterizations of the "asymptotic regression model" which, among other things, calculates various measures of nonlinear behavior, (3) a FORTRAN listing of the subroutine EVAL containing the model function and its derivatives, (4) a listing of the data set being illustrated, and (5) the output generated by use of the calling program for this illustrative problem.

Most of the output is identical to that which appeared in Ratkowsky (1983), pp. 241–242, the only difference being the inclusion here of the results of using the Hougaard (1985) skewness measure. The g_1 statistic tells us that parameter 2 is very close-to-linear ($g_1 = 0.051$), that parameter 3 exhibits perceptible negative skewness ($g_1 = -0.353$), and that parameter 1 reveals substantial positive skewness ($g_1 = 0.717$). The simulation study reported in Ratkowsky (1983), pp. 242–245, for this problem, confirms these indications.

It may also be appropriate to report here an error that appeared in the code for subroutine CURVE in Ratkowsky (1983). The second line of code on p. 228 should read

```
IF(NPMIN1.LT.1) GO TO 1111
```

replacing LE with LT. I am grateful to Dr. Bryson C. Bates, CSIRO, Western Australia, for pointing out this error, which has occasionally resulted in incorrect intrinsic nonlinearities.

1 FORTRAN listing of subroutine SKEW

```
      SUBROUTINE SKEW(PARAM,VARC,A,DERIVS,WK1,AJ,VAR,NPTS,NP,
     +               ND1,LUNOUT,EVAL,W,THIRD,G1)
C
C----------------------------------------------------------------------
C
C     SUBROUTINE SKEW CALCULATES THE THIRD MOMENT AND SKEWNESS
C     COEFFICIENT FOR EACH PARAMETER SEPARATELY. THE THIRD
C     MOMENT IS BASED ON THE FORMULA OF HOUGAARD (J. ROY. STATIST.
C     SOC. B47, 103-114 (1985)).
C
C     ON ENTRY
C     (NOTE - ALL ARGUMENTS IN THE ENTRY LIST ARE RETURNED UNALTERED.)
C     PARAM   REAL(K), K.GE.NP
C             THE PARAMETER ESTIMATES FOR THE MODEL AT CONVERGENCE.
C     VARC    REAL(ND1,K), ND1.GE.NP, K.GE.NP
C             THE PARAMETER VARIANCE-COVARIANCE MATRIX.
C     A       REAL(K), K.GE.NPTS*NP*NP
C             USED TO STORE THE SECOND-ORDER PARTIAL DERIVATIVES.
C     DERIVS  REAL(K), K.GT.NPTS*NP
C             USED TO STORE THE FIRST-ORDER PARTIAL DERIVATIVES.
C     WK1     REAL(K), K.GE.NP*(NP+1)/2
C             A GENERAL WORKING VECTOR.
C     AJ      REAL(K), K.GE.NP*NP
C             USED TO STORE INTERMEDIATE RESULTS.
C     VAR     REAL
C             THE RESIDUAL VARIANCE FOR THE MODEL AT CONVERGENCE.
C     NPTS    INTEGER
C             THE NUMBER OF OBSERVATIONS.
C     NP      INTEGER
C             THE NUMBER OF PARAMETERS.
C     ND1     INTEGER
C             THE LEADING DIMENSION OF THE MATRIX VARC.
C     LUNOUT  INTEGER
C             THE LOGICAL UNIT NUMBER TO WHICH ERROR MESSAGES AND RESULTS
C             ARE TO BE PRINTED.
C     EVAL    SUBROUTINE
C             A USER-SUPPLIED SUBROUTINE (THE SAME AS CALLED BY SOLVE AND
C             BATES, ETC.). EVAL MUST BE DECLARED EXTERNAL IN THE CALLING
C             PROGRAM.
C     W       REAL(K), K.GE.NP*NP*NP
C             USED TO STORE INTERMEDIATE RESULTS.
C     ON EXIT
C     THIRD   REAL(K), K.GE.NP
C             THE THIRD MOMENT OF HOUGAARD (1985).
C     G1      REAL(K), K.GE.NP
C             THE SKEWNESS COEFFICIENT (I.E., THE STANDARDIZED THIRD
C             MOMENT).
C
C     SUBROUTINES AND FUNCTIONS REQUIRED
C        USER-SUPPLIED        EVAL
C
C     HISTORY
C        JANUARY 1988 PROGRAMMED (D.A. RATKOWSKY)
C
C----------------------------------------------------------------------
C
```

```
      EXTERNAL EVAL
      INTEGER I,J,K,IFAIL,ITASK,NP,NPTS,ND1,NW,LUNOUT
      REAL PARAM(1),A(NPTS,NP,1),DERIVS(NPTS,1),
     + AJ(NP,1),WK1(1),VARC(ND1,1),W(NP,NP,1),THIRD(1),G1(1),WW,
     + ACCUM,ZERO,CONS,VAR
C
      DATA ZERO/0.0E0/,CONS/1.5E0/
C
C   CALCULATE FIRST AND SECOND DERIVATIVES
      DO 100 I=1,NPTS
        ITASK=1
        CALL EVAL(PARAM,I,ITASK,WK1,IFAIL)
        IF(IFAIL.NE.0) GO TO 1111
        DO 200 J=1,NP
          DERIVS(I,J)=WK1(J)
  200   CONTINUE
        ITASK=2
        CALL EVAL(PARAM,I,ITASK,WK1,IFAIL)
        IF(IFAIL.NE.0) GO TO 1111
        NW=0
        DO 300 J=1,NP
          DO 400 K=1,J
            NW=NW+1
            WW=WK1(NW)
            A(I,J,K)=WW
            IF(J.NE.K) A(I,K,J)=WW
  400     CONTINUE
  300   CONTINUE
  100 CONTINUE
C
C   MULTIPLY SECOND DERIVATIVES BY TRANSPOSE OF FIRST DERIVATIVES
      DO 500 I=1,NP
        DO 600 J=1,I
          DO 700 K=1,NP
            ACCUM=ZERO
            DO 800 L=1,NPTS
              ACCUM=ACCUM + DERIVS(L,K) * A(L,I,J)
  800       CONTINUE
            W(I,J,K)=ACCUM
            IF(I.NE.J) W(J,I,K)=ACCUM
  700     CONTINUE
  600   CONTINUE
  500 CONTINUE
C
C INVERSE OF HOUGAARD'S J MATRIX
      DO 900 I=1,NP
        DO 1000 J=1,I
          AJ(I,J)=VARC(I,J)/VAR
          IF(I.NE.J) AJ(J,I)=AJ(I,J)
 1000   CONTINUE
  900 CONTINUE
C
C   CALCULATE 'ASYMPTOTIC' THIRD MOMENT OF HOUGAARD
      DO 1100 I=1,NP
        ACCUM=ZERO
```

```
         DO 1200 J=1,NP
           DO 1300 K=1,NP
             DO 1400 L=1,NP
               ACCUM=ACCUM + AJ(I,J) * AJ(I,K) * AJ(I,L) * (W(J,K,L)
     +             + W(K,J,L) + W(L,J,K))
 1400         CONTINUE
 1300       CONTINUE
 1200     CONTINUE
         THIRD(I)=-VAR*VAR*ACCUM
         G1(I)=THIRD(I)/VARC(I,I)**CONS
 1100 CONTINUE
C
      WRITE(LUNOUT,10)
      WRITE(LUNOUT,20) (I,THIRD(I),G1(I),I=1,NP)
      RETURN
C
 1111 CONTINUE
C     USER-FLAGGED ERROR IN EVAL.
      IFAIL=8
      WRITE(LUNOUT,30)
      RETURN
C
C  FORMAT STATEMENTS
C
   10 FORMAT(//11X,44H'ASYMPTOTIC' THIRD MOMENT OF HOUGAARD (1985)/
     + 7X,9HPARAMETER,5X,12HTHIRD MOMENT,8X,12HG1 STATISTIC  /)
   20 FORMAT(8X,I4,1X,E20.8,F20.6)
   30 FORMAT(//1H ,41H*** SUBROUTINE SKEW    USER-FLAGGED ERROR,
     +19H IN SUBROUTINE EVAL//)
      END
```

2 FORTRAN listing of calling program for fitting the asymptotic regression model, calculating nonlinearity measures, etc.

```
C
C
C     MAXIMUM OF 9 PARAMETERS, 75 DATA POINTS, 5 X-VARIATES
C
      EXTERNAL EVAL
C
      INTEGER JPVT(9),LUNIN,LUNOUT,NXVAR,NPTS,NP,NITS,ND1,I,JOB,
     +        IFAIL,ITIT(80)
      REAL Y(75),PARAM(9),VARC(9,9),DERIVS(675),A(6075),
     +     QRAUX(9),WK1(45),WK2(75),DIFF(75),BIAS(9),X,W(729),
     +     AJ(81),ALPHA,RSS,VAR,FLOAT
C
      COMMON X(75,5)
C
C     VALUES FOR CONSTANTS.
C
```

```
      DATA LUNIN/5/,LUNOUT/6/,NITS/30/,ND1/9/
      DATA ALPHA/1.0E-6/
C
C     READ AND PRINT THE X- AND Y-VALUES.
C
      READ(LUNIN,10)  (ITIT(I),I=1,80)
      WRITE(LUNOUT,10)  (ITIT(I),I=1,80)
      READ(LUNIN,*)  NPTS,NP,NXVAR
      READ(LUNIN,*)  (PARAM(I),I=1,NP)
      DO 100 I=1,NPTS
          READ(LUNIN,*)  (X(I,J),J=1,NXVAR),Y(I)
          WRITE(LUNOUT,30)  (X(I,J),J=1,NXVAR),Y(I)
  100 CONTINUE
C
C  A. FIND THE LEAST-SQUARES PARAMETER ESTIMATES.
C
      JOB=11111
      CALL SOLVE(Y,PARAM,ALPHA,NITS,NPTS,NP,ND1,LUNOUT,JOB,EVAL,
     +          VARC,RSS,IFAIL,DERIVS,QRAUX,JPVT,DIFF,WK1)
C
C  B. CALCULATE PARAMETER BIASES AND CURVATURE MEASURES.
C
      JOB=111
      VAR=RSS/FLOAT(NPTS-NP)
      CALL SKEW(PARAM,VARC,A,DERIVS,WK1,AJ,VAR,NPTS,NP,
     +          ND1,LUNOUT,EVAL,W,BIAS,QRAUX)
      CALL BATES(PARAM,VAR,NITS,NPTS,NP,LUNOUT,JOB,EVAL,BIAS,IFAIL,
     +          A,DERIVS,VARC,QRAUX,WK1,WK2,JPVT)
      STOP
C
C     FORMAT STATEMENTS.
C
   10 FORMAT(80A1)
   30 FORMAT(8F15.6)
      END
```

3 FORTRAN listing of SUBROUTINE EVAL containing the code for evaluating the parameterization of the asymptotic regression model and its derivatives.

```
      SUBROUTINE EVAL(PARAM,I,ITASK,WK1,IFAIL)
C
C     EXAMPLE PROGRAM FOR THE MODEL
C
C
C               E(Y) = PARAM(1) - PARAM(2) * PARAM(3) ** X
      INTEGER I,ITASK,IFAIL
      REAL PARAM(3),WK1(6),X,XX,ZERO,ONE,TWO
      REAL P1,P2,P3
C
      COMMON X(75,5)
C
      DATA ZERO/0.0E0/,ONE/1.0E0/,TWO/2.0E0/
```

```
C
      IFAIL=0
      XX=X(I,1)
      P1=PARAM(1)
      P2=PARAM(2)
      P3=PARAM(3)
C
      IF(ITASK.NE.0) GO TO 1111
C          PREDICTED VALUE IS REQUIRED.
      WK1(1)=P1-P2*P3**XX
      RETURN
C
 1111 CONTINUE
      IF(ITASK.NE.1) GO TO 2222
C          FIRST DERIVATIVES REQUIRED.
      WK1(1)=ONE
      WK1(2)=-P3**XX
      WK1(3)=-P2*XX*P3**(XX-ONE)
      RETURN
C
 2222 CONTINUE
      IF(ITASK.NE.2) GO TO 3333
C          SECOND DERIVATIVES REQUIRED.
      WK1(1)=ZERO
      WK1(2)=ZERO
      WK1(3)=ZERO
      WK1(4)=ZERO
      WK1(5)=-XX*P3**(XX-ONE)
      WK1(6)=-P2*XX*(XX-ONE)*P3**(XX-TWO)
      RETURN
C
 3333 CONTINUE
C      ERROR - SET IFAIL TO A NONZERO VALUE.
      IFAIL=1
      RETURN
      END
```

4   Leaf Production Data Set (Ratkowsky, 1983, p. 102)

```
LEAF PRODUCTION DATA
6 3 1
.3 .3 .98
12 .094
23 .119
40 .199
92 .260
156 .309
215 .331
```

5   Output from above problem

LEAF PRODUCTION DATA

```
        12.000000        0.094000
        23.000000        0.119000
        40.000000        0.199000
        92.000000        0.260000
       156.000000        0.309000
       215.000000        0.331000
```

                            PARAMETER ESTIMATES
```
          RSS             1                2                3
  0.373183e-02 0.300000e+00 0.300000e+00 0.980000e+00
  0.620509e-03 0.331851e+00 0.292823e+00 0.983632e+00
  0.605641e-03 0.334930e+00 0.295523e+00 0.983811e+00
  0.605635e-03 0.335012e+00 0.295514e+00 0.983828e+00
  0.605635e-03 0.335020e+00 0.295512e+00 0.983829e+00
  0.605635e-03 0.335020e+00 0.295512e+00 0.983830e+00
  0.605635e-03 0.335020e+00 0.295512e+00 0.983830e+00
```

                    PARAMETER ESTIMATES AT CONVERGENCE

```
PARAMETER     ESTIMATE             SE              T
    1       0.335020e+00  0.173160e-01        19.35
    2       0.295512e+00  0.195638e-01        15.11
    3       0.983830e+00  0.370336e-02       265.66
```

                    PARAMETER VARIANCE-COVARIANCE MATRIX

```
    1   0.299845e-03
    2   0.922989e-04  0.382742e-03
    3   0.535329e-04  -.136012e-04  0.137148e-04
```

                        PARAMETER CORRELATIONS

```
    1   1.000000
    2   0.272455     1.000000
    3   0.834789    -0.187729     1.000000
```

```
UNIT              Y             FITTED          RESIDUAL
   1     0.940000e-01    0.920159e-01     0.198406e-02
   2     0.119000e+00    0.131909e+00    -0.129094e-01
   3     0.199000e+00    0.181073e+00     0.179269e-01
   4     0.260000e+00    0.269071e+00    -0.907140e-02
   5     0.309000e+00    0.311789e+00    -0.278862e-02
   6     0.331000e+00    0.326142e+00     0.485843e-02
```

```
      'ASYMPTOTIC' THIRD MOMENT OF HOUGAARD (1985)
PARAMETER     THIRD MOMENT        G1 STATISTIC

    1      0.37252547e-05          0.717483
    2      0.38332459e-06          0.051193
    3     -0.17944572e-07         -0.353302
```

ACCELERATION ARRAY (PARAMETER-EFFECTS PORTION)

```
        0.1504       0.4049      -0.0321
        0.4049       1.0903      -0.0865
       -0.0321      -0.0865      -0.0806


       -0.0445      -0.1198      -0.0244
       -0.1198      -0.3224      -0.0657
       -0.0244      -0.0657      -0.0134


        0.0981       0.2641       0.0538
        0.2641       0.7110       0.1450
        0.0538       0.1450       0.0296
```

BOX'S BIAS

PARAMETER	LS ESTIMATE	BIAS	PERCENT BIAS
PARAMETER 1	0.33502050e+00	0.29713907e-02	0.8869
PARAMETER 2	0.29551209e+00	0.47364067e-02	1.6028
PARAMETER 3	0.98382970e+00	-0.21805994e-03	-0.0222

MAXIMUM CURVATURE MEASURES

INTRINSIC (IN)	0.2318
PARAMETER-EFFECTS (PE)	1.5264

Author Index

Amari, S. I., 26

Bacon, D. W., 120
Bacon, G. E., 166
Bard, Y., 21
Bates, D. M., 18, 23–25, 27,
    29, 59–62, 199–201, 207
Beale, E. M. L., 199
Bělehrádek, J., 91, 107
Bellman, R. E., 45
Belsley, D. A., 17
Bleasdale, J. K. A., 46, 103
Box, M. J., 24, 26–27, 29, 207
Bradley, R. S., 51, 92

Bragg, R. H., 165
Bruin, S., 51
Brunauer, S., 51

Campbell, N. A., 114
Causton, D. R., 47
Chambers, J., 21
Chen, C. S., 52
Chirife, J., 51–53, 58
Clarke, G. P. Y., 26
Cook, R. D., 17, 25–26, 38
Cox, D. R., 26

Subject Index